Springer Theses

Recognizing Outstanding Ph.D. Research

Recognizing Outstanding Ph.D. Research

Aims and Scope

The series "Springer Theses" brings together a selection of the very best Ph.D. theses from around the world and across the physical sciences. Nominated and endorsed by two recognized specialists, each published volume has been selected for its scientific excellence and the high impact of its contents for the pertinent field of research. For greater accessibility to non-specialists, the published versions include an extended introduction, as well as a foreword by the student's supervisor explaining the special relevance of the work for the field. As a whole, the series will provide a valuable resource both for newcomers to the research fields described, and for other scientists seeking detailed background information on special questions. Finally, it provides an accredited documentation of the valuable contributions made by today's younger generation of scientists.

Theses are accepted into the series by invited nomination only and must fulfill all of the following criteria

- They must be written in good English.
- The topic should fall within the confines of Chemistry, Physics, Earth Sciences, Engineering and related interdisciplinary fields such as Materials, Nanoscience, Chemical Engineering, Complex Systems and Biophysics.
- The work reported in the thesis must represent a significant scientific advance.
- If the thesis includes previously published material, permission to reproduce this must be gained from the respective copyright holder.
- They must have been examined and passed during the 12 months prior to nomination.
- Each thesis should include a foreword by the supervisor outlining the significance of its content.
- The theses should have a clearly defined structure including an introduction accessible to scientists not expert in that particular field.

More information about this series at http://www.springer.com/series/8790

Xinlu Xu

Phase Space Dynamics in Plasma Based Wakefield Acceleration

Doctoral Thesis accepted by
Tsinghua University, Beijing, China

 Springer

Author
Dr. Xinlu Xu
Department of Engineering Physics
Tsinghua University
Beijing, China

Supervisor
Prof. Wei Lu
Department of Engineering Physics
Tsinghua University
Beijing, China

ISSN 2190-5053   ISSN 2190-5061   (electronic)
Springer Theses
ISBN 978-981-15-2383-0   ISBN 978-981-15-2381-6   (eBook)
https://doi.org/10.1007/978-981-15-2381-6

This Springer imprint is published by the registered company Springer Nature Singapore Pte Ltd.
The registered company address is: 152 Beach Road, #21-01/04 Gateway East, Singapore 189721, Singapore

Supervisor's Foreword

The two most important long-term applications of plasma accelerators are a linear collider operating at the energy frontier of particle physics and a fifth-generation X-ray free-electron-laser (X-FEL). Critical to both of these applications is the ability to generate an electron beam with excellent beam quality (low emittance, high current and low energy spread) in plasma and couple this beam into another plasma section or into a conventional accelerator element. In this dissertation, Xinlu addresses fundamental yet unaddressed topics in plasma wakefield acceleration with respect to obtaining high-quality electron beams which could have applications for the above two important facilities. Xinlu's dissertation includes very original research that addresses these issues as well as an analysis of a numerical instability that can arise when simulating plasma-based acceleration. His dissertation research is outstanding. It is original, profound, and of broad interest. He has a clear grasp of the important issues in this field, of the current status of research, of experimental realities, and of advanced simulation techniques.

Plasma wave accelerating structures are very compact (high frequency and short wavelength) so synchronously injecting particles into them is a challenge. One method that has received much recent attention is the ionization injection of electrons. In Chap. 2, Xinlu uses original theory and advanced simulations to study the complicated phase space dynamics that is involved when electrons are injected and accelerated in nonlinear plasma waves. He finds that the final emittance depends on the duration of the injection and that the final emittance depends on the square of the initial radius. The dynamics of the projected emittance includes phase mixing of transverse slices with different betatron phases (i.e., transverse phase mixing) and the projected and slice emittances are affected by particles injected at different times being in the same slice (i.e., longitudinal phase mixing). This work demonstrates that the final emittance depends on the above dynamics and not the initial thermal emittance. He also uses the results to propose a novel two-color laser accelerator scheme that uses a longer wavelength laser to make a wake and a shorter wavelength laser to ionize additional electrons that are more easily trapped by the wake. It is worth mentioning that a better understanding of the longitudinal mixing effect has led to new results that show if the ionization laser is located where there is a slope in the wake potential then the final beam can be bunched in real space. This interesting phenomenon shows the complication of the physics and the future impact of his research.

In Chap. 3, Xinlu also provides a clever solution to a little appreciated issue for plasma-based acceleration. The focusing forces in plasma waves are proportional to the plasma density and they are orders of magnitude higher than in conventional optics. Therefore, when transporting electron bunches from one plasma section to another or from a plasma section to a conventional accelerator or beam optics section there can be a beta function mismatch. Therefore, a matched beam in one section will be far from matched in another section. If there is energy spread on the beam transporting a beam between two sections with very different beta functions can lead to substantial emittance growth. In his dissertation, Xinlu studies an original idea in which a properly chosen density profile at the end (or beginning) of a plasma section can be used to rotate the beam in transverse phase space so that it is matched into the next section. In Chap. 4, Xinlu uses the results in Chap. 3 to offer a possible path forward for a self-consistent start-to-end design of a compact X-FEL based on plasma-based acceleration. In the very near future, this design may be tested in fully nonlinear particle-based simulations. Furthermore, due to his close connection to experimentalists, his design is based on experimentally reasonable parameters. Therefore, the ideas from this chapter are extremely important.

Last in Chap. 5, Xinlu addresses a numerical instability that makes studying low emittance beams over long regions of plasma challenging using particle-in-cell (PIC) techniques. PIC simulations are currently the most accurate method to study and design plasma-based acceleration stages. Such simulations require many CPU hours and need to be run on the world's largest computers. Recently, researchers have explored the possibility of carrying simulations in a Lorentz-boosted frame in which a Lorentz-contracted plasma is sent towards the driver. As long as there is no reflected light then the CPU hours needed to model a given plasma length becomes 10,000 times less. Unfortunately, when a plasma or an electron beam propagates on a grid with speeds near the speed of light a strong numerical Cerenkov instability (NCI) occurs. Therefore, the use of a Lorentz-boosted frame to simulate plasma-based acceleration has not been fully utilized. Xinlu carried out a careful analysis of the numerical dispersion relation for general PIC methods. Unlike, in a real system where there would be no coupling between transverse and longitudinal waves, his dispersion relations showed that this did occur, and it identified the region in k space where it occurred for a variety of Maxwell field solvers and deposition options. Based on this work, he and others have developed methods to avoid the NCI.

In summary, the dissertation of Xinlu is outstanding. It is full of original and important work. His results will also impact the future path of the field of plasma-based acceleration.

Beijing, China   Prof. Wei Lu
August 2019

Preface

Plasma-based accelerators have made great progress in the past decade, and start to be considered seriously as possible drivers for future compact X-ray free-electron-lasers (X-FELs) and linear colliders. For these challenging applications, how to generate electron beams with extremely high quality in plasma accelerators and how to transport these beams without quality degradation become crucial. In this dissertation, we systematically studied several key issues of the beam phase space dynamics in plasma accelerators through theoretical analysis and particle-in-cell (PIC) simulation, and the major topics include the phase space dynamics of the ionization-based injection methods, the phase space matching between accelerator stages, and a preliminary design study of X-FELs driven by plasma accelerators.

In Chap. 2, the evolution of beam phase space in ionization injection in plasma wakefields is studied using theoretical analysis and PIC simulations. Two key phase mixing processes, namely longitudinal and transverse phase mixing, are identified to be responsible for the complex emittance dynamics that shows initial rapid growth followed by oscillation, decay, and eventually slow growth to saturation. An analytical theory is developed to include the effects of injection distance (time), acceleration distance, wakefield structure, and nonlinear space charge forces. Formulas for the emittance in the low and high charge regimes are presented and verified through PIC simulations. A two-color laser ionization injection scheme for generating high-quality beams is also examined through PIC simulations. In the end, the intrinsic phase space discretization in laser-triggered ionization injection is analyzed.

In Chap. 3, the phase space matching between plasma acceleration stages and between plasma stages and traditional accelerator components is studied. Without proper matching, catastrophic emittance growth in the presence of finite energy spread may occur when the beam propagates through different stages and components due to the drastic differences of transverse focusing strength. A method using longitudinally tailored plasma structures as phase space matching components to properly guide the beam through stages is proposed. Theoretical analysis and particle-in-cell simulations are utilized to show clearly how these structures can work in four different scenarios.

In Chap. 4, a two-stage scheme of laser wakefield accelerator (LWFA) for driving a compact X-FEL is studied. In this scheme, an ultrashort 50 MeV electron beam with very high brightness is generated in a high density plasma using density down ramp injection, and then the beam is sent through a low density plasma accelerator for further acceleration to reach a few GeV energy and to reduce the relative slice energy spread down to $\sim 0.4\%$. Plasma density matching structures are properly designed according to Chap. 3 to maintain the beam quality between the plasma stages. Finally, the X-FEL performance driven by such a high-quality electron beam is studied using Genesis 1.3.

In Chap. 5, the numerical noise issue in boosted frame simulation of LWFA is systematically studied. A general multidimensional numerical dispersion relation for a relativistic drifting plasma in PIC simulation is derived, and based on this dispersion relation, the source of the instability and its growth rate is clearly identified. An asymptotic expression that permits rapid parameter scans for the ranges of unstable modes is also derived, and this asymptotic expression confirms and explains the existence of a 'magic time step' in boosted frame simulations. To significantly reduce the numerical noise, a spectral solver is adopted in a new PIC code UPIC-EMMA, and very good agreement between UPIC-EMMA boosted frame simulations and OSIRIS lab frame simulations is obtained.

August 2019   Xinlu Xu
SLAC National Accelerator Laboratory
Menlo Park, CA, USA

Acknowledgements

First of all, I would like to thank my Ph.D. advisor, Prof. Wei Lu, for all the knowledge, support and guidance over the course of my study at Tsinghua University. I'll never forget the good old times when we brainstormed together, when we polished manuscripts word by word, and when we chatted about interesting things in life. I would also like to express my gratitude to Prof. Chuangxiang Tang for his continuous support and encouragement during the past years.

I would also like to thank my very good friend, and collaborator, Peicheng Yu. Our collaboration starts since I was at UCLA as a visiting student and continues after I went back to Tsinghua. His tastes on physics and his attitudes on life impact me a lot.

I am grateful to Prof. Huaibi Chen, Prof. Wenhui Huang, Prof. Qingzi Xing, Prof. Jianfei Hua, Prof. Lixian Yan and Prof. Yingchao Du for their help. A special thanks to Prof. Warren Mori and Prof. Chan Joshi for their guidance when I was at UCLA. I would also like to thank my valued colleagues, including Dr. Weiming An, Dr. Jiaru Shi, Dr. Renkai Li, Dr. Houjun Qian, Dr. Jiaqi Qiu, Dai Wu, Hao Zha, Haisheng Xu, Chen Li, for not only the academic collaborations, but also for their friendships and companions for the past years.

Lastly, I would like to thank my parents and my girlfriend for their love and support.

Contents

Chapter 1
Introduction

Abstract Morden accelerators use radio-frequency electromagnetic waves to accelerate charged particles and have been the tool of choice for unraveling the structure of matter, space, and time. However the capability of existing accelerator technology has plateaued due to its limited acceleration gradient $10 \sim 100$ MV/m, so that a future accelerator at the energy frontier will be so large and expensive that it is not clear it will be built. Plasma based acceleration can sustain ultra high acceleration gradient $10 \sim 100$ GV/m due to the ionized medium—plasma, thus much shrink the size and reduce the cost of high energy accelerators. In this chapter we will briefly introduce the basic concept of plasma based acceleration and its history.

1.1 Plasma Based Acceleration

In physics, the ultimate structure of matter and the unification of the force of the nature have always been one of the most noble goals [1]. Since early 20th century, experiments based on high energy particles collisions have become the major tool to reveal the mystery of the sub-nuclear world. In 1910, by bombarding a thin metal foil with 10 MeV alpha particles, Rutherford discovered an atom is composed of a tiny nucleus surrounded by electrons at a much greater distance. The alpha particles used by Rutherford are from radioactive decay and their energy is limited to the MeV-level. To generate higher energy particles, many generations of accelerators have been proposed and built in the past century [2]. In 2012, proton collisions with energy up to a few TeV at the Large Hadron Collider (LHC) [3] at CERN confirm the existence of the Higgs Boson particle [4], which marks the latest great success of high energy physics based on particle accelerators. To generate the multi-TeV protons needed, a 27 km long tunnel was built. To further refined study of the Higgs Boson and the standard model, people now are proposing a new machine called International Linear Collider (ILC) [5] to carry out this task. In this machine electron and positron will collide at 500 GeV energy, and this needs a 31 km long machine based on current radio-frequency (RF) technology [6] at a cost more than 20 billion dollars. The huge size of LHC and ILC are approaching the limit of modern engineering, and this gives strong impetus for finding out new ways to accelerate particles at much higher gradient.

© Springer Nature Singapore Pte Ltd. 2020

X. Xu, *Phase Space Dynamics in Plasma Based Wakefield Acceleration*,

Springer Theses, https://doi.org/10.1007/978-981-15-2381-6_1

Indeed over the past few decades, many new acceleration concepts have been proposed and are currently under active testing [7–12]. Among these schemes, plasma based concept [13] has achieved astonishing progress worldwide and is becoming one of the leading candidates for the future machines. In the plasma based acceleration, an intense driver, a laser pulse or charged particle beam, is utilized to excite a wake in a plasma. Since it is fully ionized, the plasma medium is free of breakdown and can sustain gradient on the order of GV/cm, which is more than three orders of magnitude higher than the acceleration gradient in the RF-based technologies, thus it may significantly reduce the size and the cost of the future high energy accelerators.

In 1979, Tajima and Dawson proposed to use an intense short pulse laser to create a wake of plasma oscillations for accelerating electrons [13], and this marks the beginning of the field of plasma based acceleration. In this concept, as the intense laser pulse propagates in the plasma, the plasma electrons are displaced away from the ion background by the ponderomotive force of the laser pulse. After the laser passing by, the electrons are pulled back by the space charge force from the ion background and oscillate around their equilibrium positions. Such oscillations have a phase velocity close to the group velocity of the laser pulse, and it also has a longitudinal component of the electric field. These two features are ideal for accelerating charged particles to high energy. The typical acceleration gradient of this plasma oscillation is on the order of the non-relativistic wave breaking limit $E_{wb} = mc\omega_p/e$ [14], where ω_p is the plasma frequency. In a typical plasma of $10^{18}\,\text{cm}^{-3}$ density, $E_{wb} = 96\,\text{GV/m}$, which is three or four orders of magnitude larger than that of the current RF technologies. Similar wakes can also be generated by charged particle beams, e.g., electrons [15, 16], positrons [17] and protons [18], where the space charge force of the beams plays a similar role of the ponderomotive force of the lasers.

The basic scheme of laser wakefield acceleration needs a laser pulse length shorter than the plasma wavelength and a laser vector potential on the order of 1. Such ultra-short high power laser did not exist in 1980s and 1990s. During that period of time, wakefield excitation and acceleration based on plasma beat wave accelerator (PBWA) [19–22] and self-modulated laser wake field accelerator (SMLWFA) [23–26] were explored extensively using relatively long laser pulses, and many milestones have been achieved, such as the first observation of the laser plasma wakes using Thomson scattering [20], the first demonstration of acceleration with external injected electrons [21] and the first acceleration with self-injected electrons in the plasma [25]. Since 2000, high power laser technology based on chirped-pulse-amplification (CPA) [27, 28] and Ti:Sapphire gain medium [29–31] has been improved significantly, enabling multi TW ($10 \sim 100\,\text{TW}$) ultra-short ($30 \sim 50\,\text{fs}$) laser system on a tabletop. With these lasers, many great new milestones have been achieved in the past decade [32–42]. In 2004, three groups demonstrated the generation of self-injected mono-energetic electron beams in the range of $100 \sim 200\,\text{MeV}$ [32–34]. Since then, many groups around the world have demonstrated similar physics and extend the energy reach to the GeV level [35–42]. Now a new record of energy 4.2 GeV has been reported very recently [43].

Parallel to the progress of the laser driven wakes, beam driven plasma wakefield acceleration [15] has also made great strides in the past decades. A series of milestone experiments were performed at Stanford linear accelerator center (SLAC) using the unique $20 \sim 40\,\text{GeV}$ electron and positron beams [17, 44–48]. The most remarkable result was the energy doubling of $42\,\text{GeV}$ electron beam in less than a meter long plasma [47]. Very recently, high efficiency acceleration ($20\% \sim 50\%$) with low energy spread has just been reported [49]. These great results show great potentials of plasma wakefield accelerators for future high energy accelerators.

It turns out most of the achievements in the past decade occurs in a special regime of wakefield excitation, the so called blowout regime. In this regime, the drivers (laser pulses or charged beams) are tightly focused to the order of the plasma skin depth with high intensity, the transverse ponderomotive force or the space charge field of the drivers are intense enough to expel the plasma electrons out of their path radially, forming a cavity with only ions. This phenomenon was first discovered in early 1990s at UCLA for both laser and particle beam drivers [50, 51]. It was also found that the field structure in such a cavity has two very nice features for acceleration and focusing of electron beams: the acceleration field is uniform on every cross section within the channel, and the transverse focusing field is uniformly linear. These two properties can help to reduce the energy spread of the electron beam during the acceleration and to conserve the emittance of the beam, making it much easier to obtain high quality acceleration. Indeed these are the major physical reasons behind the great results obtained in the past decade.

In the blowout regime, the plasma oscillations are complicated due to their intrinsic multi-dimensional and highly nonlinear nature. Further more, the motion of the electrons is relativistic and the fields in the wake are electromagnetic in nature. In 2006, a self-consistent nonlinear theoretical model for relativistic plasma wakefields in the blowout regime was developed at UCLA [52, 53]. This model can describe the shape of the ion channel driven by relativistic electron beams or intense short lasers with matched spot sizes. Based on this model, theories on the nonlinear beam loading [54, 55] and electron hosing instability [56] were also developed. In 2007, by taking into account the major physical processes that affect the laser wakefield acceleration in the blowout regime, a phenomenology theoretical framework was developed to help design a high quality stable laser wakefield accelerator [57]. In this framework, designing formulas are presented to estimate the laser and the plasma parameters and the charge and the energy of the accelerated electrons for various scenarios. These works have established a solid base for the further study of plasma based wakefield acceleration in the blowout regime.

1.2 Particle-in-Cell Simulations

Most plasma physics involves a large number of degree of freedom and are often highly nonlinear. To understand the very details of the relevant physics, numerical simulation becomes an essential and powerful tool, and this is especially true for the study of plasma wakefield acceleration due to its highly nonlinear and kinetic nature. One of the most important numerical methods used in studying the plasma wakefield acceleration is the particle-in-cell (PIC) simulations [58], which computes the motions of a collection of charged particles under their self and externally applied fields.

The basic step of PIC simulations is simple: the space is divided into grids with a resolution fine enough to resolve the shortest spatial scales of interest. The electromagnetic fields are defined and computed on these grids. At each time step, the particles are updated with new positions and momenta under the action of the electromagnetic field. The charges and current densities then are interpolated onto the grids from the new positions and velocities of particles. The fields are updated using the Maxwell equations with the new charges and current densities. The simulations can be run for the required number of time steps. Since the PIC codes work at a fundamental level with very few approximations, it has become the most important tool in plasma physics to deal with highly nonlinear and kinetic problems ever since the pioneering work of Dawson [59] and others in the early 1960s. With the fast growth of the computing power during the past decade, PIC codes have been used more and more in complicated multi-dimensional problems. Problems relevant to plasma wakefield acceleration in the blowout regime is a perfect example in this regard. We will extensively use PIC simulations in this dissertation.

1.3 Motivation and Outline

As mentioned earlier, plasma based wakefield acceleration has made great progresses in the past decade, and starts to be considered seriously as possible drivers for future compact X-ray free-electron-lasers (X-FELs) and linear colliders. For these challenging applications, how to generate electron beams with extremely high quality in plasma accelerators and how to transport these beams without quality degradation become crucial. The beam quality are frequently quantified using three parameters, i.e., the current I, the normalized emittance ϵ_n and the energy spread σ_E. Highly demanding applications like FELs and colliders always impose very stringent requirements on the beam quality. For example, to achieve high gain in FELs, the normalized emittance of the beam must satisfy $\epsilon_n < \langle \gamma_b \rangle \lambda_r$ and the relative energy spread needs to be less than the FEL parameter ρ, where $\langle \gamma_b \rangle$ is the mean relativistic factor of the beam and λ_r is the resonant wavelength; for colliders, high current and very low emittance are needed to achieve high luminosity for increasing the event rate of new particles generation. The beam quality obtained in plasma accelerators so far are still far off from these challenging requirements. Table 1.1 shows the qualities of the beam generated from recent experiments since 2010. Although the beam quality here is already suitable for certain very useful applications, such as betatron sources [46, 60–66] and Thomson scattering X-ray sources [67–70], it is still not good enough for FELs and colliders.

For high quality beam generation, the injection, acceleration and subsequent transport of the electrons must be understood thoroughly and controlled precisely. Currently, there is a lack of understanding of the 6D phase space dynamics of the trapped electrons in all the injection schemes. The key questions need to be answered include: what physics determines the beam quality in these injection schemes? How to gener-

Table 1.1 The beam parameters generated from plasma based accelerators in recent five years

Group	Charge (pC)	Divergence (mrad)	Energy and energy spread (MeV)	Injection scheme
UCLA, USA [71]	Several pC	6.3/3	92, 60%	Ionization injection
University of Michigan, USA [72]	Hundreds of pC	2.9 ± 0.8	$\sim$120, broadband	Ionization injection
Institute for radiation physics, Germany [73]	30	$10 \sim 20$	40, 7^{rms}	Self-injection
UCLA, USA [38]	3.8	4.4^{rms}	$300 \sim 1450$	Ionization injection
University of Strathclyde, UK [74]	Several pC	$\epsilon_n = 2.2 \pm 0.7 \, \mu m$	125 ± 3	Self-injection
The Blackett Lab, UK [62]	$100 \sim 300$	$1.5/1.8^{rms}$	$\sim$230, $25\% \pm 10\%^{FWHM}$	Self-injection
SIOFM, China [39]	3.7	2.6	$\sim$800, 25%	Ionization injection
LLNL, USA [40]	35	2.3	460, $< 5\%^{FWHM}$	Ionization injection
Ecole Polytechnique, France [75]	$15 \pm 7^{1.4 \sim 1.8 \, fs}$	6.6 ± 1.6	$\sim$84, 21 ± 17^{FWHM}	Self-injection
LBNL, USA [76]	$1 \sim 10$	2.5	341, 11%	Density downramp injection
Ecole Polytechnique, France [68]	$\sim$120	–	100, broadband	Self-injection
MPQ, Germany [77]	90 ± 30	$\sim$8	50, 5	Density downramp injection
GIST, South Korea [42]	80	3.3/4.3	$500 \sim 5000$	Self-injection
Ecole Polytechnique, France [78]	5	$< 2^{FWHM}$	$\sim$200, $50 \sim 70^{FWHM}$	Self-injection
UT Austin, USA [41]	63 ± 8	0.6^{FWHM}	2000, $6\%^{FWHM}$	Self-injection
University of Nebraska-Lincoln, USA [70]	23	10^{FWHM}	55, $22\%^{FWHM}$	Self-injection
Institute of Physics, China [79]	5	4	180, 10%	Self-injection
LBNL, USA [43]	6	0.3	4200, $6\%^{rms}$	Self-injection
Tsinghua University, China [80]	$\sim$1	A few mrad	$\sim$20, $1 \sim 2\%^{rms}$	Self-injection

ate high quality electron beam with proper control? What is the fundamental limit on the beam quality from these schemes? How to conserve the quality when the beam propagates between different components? Is it possible to drive a X-ray FEL using plasma based accelerators? In this dissertation, we aim at using the language and tools of accelerator physics to systematically study these issues, and the following is an outline of each chapter.

In Chap. 2, the evolution of beam phase space in ionization injection in plasma wakefields is studied using theoretical analysis and PIC simulations. Two key phase mixing processes, namely longitudinal and transverse phase mixing, are identified to be responsible for the complex emittance dynamics that shows initial rapid growth followed by oscillation, decay, and eventually slow growth to saturation. An analytic theory is developed to include the effects of injection distance (time), acceleration distance, wakefield structure, and nonlinear space charge forces. Formulas for the emittance in the low and high charge regimes are presented and verified through PIC simulations. A two-color laser ionization injection scheme for generating high quality beams is also examined through PIC simulations. In the end, the intrinsic phase space discretization in laser triggered ionization injection is analyzed.

In Chap. 3, the phase space matching between plasma acceleration stages and between plasma stages and traditional accelerator components is studied. Without proper matching, catastrophic emittance growth in the presence of finite energy spread may occur when the beam propagates through different stages and com-ponents due to the drastic differences of transverse focusing strength. A method using longitudinally tailored plasma structures as phase space matching components to properly guide the beam through stages is proposed. Theoretical analysis and particle-in-cell simulations are utilized to show clearly how these structures can work in four different scenarios.

In Chap. 4, a two-stage scheme of LWFA for driving a compact X-FEL is studied. In this scheme, an ultrashort 50 MeV electron beam with very high brightness is generated in a high density plasma using density downramp injection, and then the beam is sent through a low density plasma accelerator for further acceleration to reach a few GeV energy and to reduce the relative slice energy spread down to $\sim$0.4%. Plasma density matching structures are properly designed according to Chap. 3 to maintain the beam quality between the plasma stages. Finally, the X-FEL performance driven by such a high quality electron beam is studied using Genesis 1.3.

In Chap. 5, the numerical noise issue in boosted frame simulation of LWFA is sys-tematically studied. A general multi-dimensional numerical dispersion relation for a relativistic drifting plasma in PIC simulation is derived, and based on this dispersion relation, the source of the instability and its growth rate is clearly identified. An asymptotic expression that permits rapid parameter scans for the ranges of unstable modes is also derived, and this asymptotic expression confirms and explains the exis-tence of a 'magic time step' in boosted frame simulations. To significantly reduce the numerical noise, a spectral solver is adopted in a new PIC code UPIC-EMMA, and very good agreement between UPIC-EMMA boosted frame simulations and OSIRIS lab frame simulations is obtained.

In Chap. 6, we provide a summary and a discussion of possible future research.

References

1. Kane G (2001) Supersymmetry: unveiling of the ultimate laws of nature. Basic Books, New York City
2. Sessler A, Wilson E (2007) Engines of discovery: a century of particle accelerators. World Scientific, Hackensack
3. http://home.web.cern.ch/topics/large-hadron-collider
4. Aad G, Abajyan T, Abbott B et al (2012) Observation of a new particle in the search for the Standard Model Higgs boson with the ATLAS detector at the LHC. Phys Lett B 716(1):1–29
5. http://www.linearcollider.org/
6. Wangler TP (1998) Principles of RF linear accelerators. Wiley, New York
7. Palmer R (1972) Interaction of relativistic particles and free electromagnetic waves in presence of a static helical magnet. J Appl Phys 43(7):3014–3023
8. Courant E, Pellegrini C, Zakowicz W (1985) High-energy inverse free-electron-laser accelerator. Phys Rev A 32(5):2813–2823
9. Wernick I, Marshall T (1992) Experimental test of the inverse free-electron-laser accelerator principle. Phys Rev A 46(6):3566–3568
10. Keinigs R, Jones M, Gai W (1989) The Cherenkov wakefield accelerator. Part Accel 24:223–229
11. Rosing M, Gai W (1990) Longitudinal- and transverse-wake-field effects in dielectric structure. Phys Rev D 42(5):1829–1834
12. Ng KY (1990) Wake fields in a dielectric-lined waveguide. Phys Rev D 42:1819–1828
13. Tajima T, Dawson JM (1979) Laser electron accelerator. Phys Rev Lett 43:267–270
14. Dawson JM (1959) Nonlinear electron oscillations in a cold plasma. Phys Rev 113:383–387
15. Chen P, Dawson JM, Huff RW et al (1985) Acceleration of electrons by the interaction of a bunched electron beam with a plasma. Phys Rev Lett 54:693–696
16. Lee S, Katsouleas T, Muggli P et al (2002) Energy doubler for a linear collider. Phys Rev ST Accel Beams 5:011001
17. Blue BE, Clayton CE, O'Connell CL et al (2003) Plasma-wakefield acceleration of an intense positron beam. Phys Rev Lett 90:214801
18. Caldwell A, Lotov K, Pukhov A et al (2009) Proton-driven plasma-wakefield acceleration. Nat Phys 5(5):363–367
19. Joshi C, Mori WB, Katsouleas T et al (1984) Ultrahigh gradient particle acceleration by intense laser-driven plasma density waves. Nature 311(5986):525–529
20. Clayton CE, Joshi C, Darrow C et al (1985) Relativistic plasma-wave excitation by collinear optical mixing. Phys Rev Lett 54:2343–2346
21. Clayton CE, Marsh KA, Dyson A et al (1993) Ultrahigh-gradient acceleration of injected electrons by laser-excited relativistic electron plasma waves. Phys Rev Lett 70:37–40
22. Everett M, Lal A, Gordon D et al (1994) Trapped electron acceleration by a laser-driven relativistic plasma-wave. Nature 368(6471):527–529
23. Sprangle P, Esarey E, Krall J et al (1992) Propagation and guiding of intense laser pulses in plasmas. Phys Rev Lett 69:2200–2203
24. Coverdale CA, Darrow CB, Decker CD et al (1995) Propagation of intense subpicosecond laser pulses through underdense plasmas. Phys Rev Lett 74:4659–4662
25. Modena A, Najmudin Z, Dangor AE et al (1995) Electron acceleration from the breaking of relativistic plasma-waves. Nature 377(6550):606–608
26. Malka V, Fritzler S, Lefebvre E et al (2002) Electron acceleration by a wake field forced by an intense ultrashort laser pulse. Science 298(5598):1596–1600
27. Strickland D, Mourou G (1985) Compression of amplified chirped optical pulses. Opt Commun 56(3):219–221
28. Maine P, Strickland D, Bado P et al (1988) Generation of ultrahigh peak power pulses by chirped pulse amplification. IEEE J Quantum Electron 24(2):398–403
29. Mourou G, Umstadter D (1992) Development and applications of compact high-intensity lasers. Phys Fluids B 4(7, 2):2315–2325

30. Joshi C, Corkum PB (1995) Interactions of ultra-intense laser light with matter. Phys Today 48(1):36–43
31. Mourou GA, Barty CPJ, Perry MD (1998) Ultrahigh-intensity lasers: physics of the extreme on a tabletop. Phys Today 51(1):22–28
32. Mangles SPD, Murphy CD, Najmudin Z et al (2004) Monoenergetic beams of relativistic electrons from intense laser-plasma interactions. Nature 431(7008):535–538
33. Geddes CGR, Toth C, Tilborg J et al (2004) High-quality electron beams from a laser wakefield accelerator using plasma-channel guiding. Nature 431(7008):538–541
34. Faure J, Glinec Y, Pukhov A et al (2004) A laser-plasma accelerator producing monoenergetic electron beams. Nature 431(7008):541–544
35. Leemans WP, Nagler B, Gonsalves AJ et al (2006) GeV electron beams from a centimetre-scale accelerator. Nat Phys 2(10):696–699
36. Hafz NAM, Jeong TM, Choi IW et al (2008) Stable generation of GeV-class electron beams from self-guided laser-plasma channels. Nat Photon 2(9):571–577
37. Kneip S, Nagel SR, Martins SF et al (2009) Near-GeV acceleration of electrons by a nonlinear plasma wave driven by a self-guided laser pulse. Phys Rev Lett 103:035002
38. Clayton CE, Ralph JE, Albert F et al (2010) Self-guided laser wakefield acceleration beyond 1 GeV using ionization-induced injection. Phys Rev Lett 105:105003
39. Liu JS, Xia CQ, Wang WT et al (2011) All-optical cascaded laser wakefield accelerator using ionization-induced injection. Phys Rev Lett 107:035001
40. Pollock BB et al (2011) Demonstration of a narrow energy spread, ~ 0.5 GeV electron beam from a two-stage laser wakefield accelerator. Phys Rev Lett 107:045001
41. Wang X, Zgadzaj R, Fazel N et al (1988) Quasi-monoenergetic laser-plasma acceleration of electrons to 2 GeV. Nat Commun 2013:4
42. Kim HT, Pae KH, Cha HJ et al (2013) Enhancement of electron energy to the multi-GeV regime by a dual-stage laser-wakefield accelerator pumped by petawatt laser pulses. Phys Rev Lett 111:165002
43. Leemans WP, Gonsalves AJ, Mao HS et al (2014) Multi-GeV electron beams from capillary-discharge-guided subpetawatt laser pulses in the self-trapping regime. Phys Rev Lett 113:245002
44. Muggli P, Blue BE, Clayton CE et al (2004) Meter-scale plasma-wakefield accelerator driven by a matched electron beam. Phys Rev Lett 93:014802
45. Hogan MJ, Barnes CD, Clayton CE et al (2005) Multi-GeV energy gain in a plasma-wakefield accelerator. Phys Rev Lett 95:054802
46. Johnson DK, Auerbach D, Blumenfeld I et al (2006) Positron production by X rays emitted by betatron motion in a plasma wiggler. Phys Rev Lett 97:175003
47. Blumenfeld I, Clayton CE, Decker FJ et al (2007) Energy doubling of 42 GeV electrons in a metre-scale plasma wakefield accelerator. Nature 445(7129):741–744
48. Vafaei-Najafabadi N, Marsh KA, Clayton CE et al (2014) Beam loading by distributed injection of electrons in a plasma wakefield accelerator. Phys Rev Lett 112:025001
49. Litos M, Adli E, An W et al (2014) High-efficiency acceleration of an electron beam in a plasma wakefield accelerator. Nature 515(7525):92–95
50. Rosenzweig J, Breizman B, Katsouleas T et al (1991) Acceleration and focusing of electrons in two-dimensional nonlinear plasma wake fields. Phys Rev A 44:R6189–92
51. Mori W, Katsouleas T, Darrow C et al (1991) Laser wakefields at UCLA and LLNL. In: Proceedings of particle accelerator conference, San Francisco, California
52. Lu W, Huang C, Zhou M et al (2006) Nonlinear theory for relativistic plasma wakefields in the blowout regime. Phys Rev Lett 96:165002
53. Lu W, Huang C, Zhou M et al (2006) A nonlinear theory for multidimensional relativistic plasma wave wakefields. Phys Plasma 13:056709
54. Tzoufras M, Lu W, Tsung FS et al (2008) Beam loading in the nonlinear regime of plasma-based acceleration. Phys Rev Lett 101:145002
55. Tzoufras M, Lu W, Tsung FS et al (2009) Beam loading by electrons in nonlinear plasma wakes. Phys Plasmas 16(5):056705

56. Huang C, Lu W, Zhou M et al (2007) Hosing instability in the blow-out regime for plasma-wakefield acceleration. Phys Rev Lett 99:255001
57. Lu W, Tzoufras M, Joshi C et al (2007) Generating multi-GeV electron bunches using single stage laser wakefield acceleration in a 3D nonlinear regime. Phys Rev ST Accel Beams 10:061301
58. Birdsall CK, Bruce LA (1985) Plasma physics via computer simulation. McGraw-Hill, New York
59. Dawson J (1962) One-dimensional plasma model. Phys Fluids 5(4):445–459
60. Németh K, Shen B, Li Y et al (2008) Laser-driven coherent betatron oscillation in a laser-wakefield cavity. Phys Rev Lett 100:095002
61. Kneip S, Nagel SR, Bellei C et al (2008) Observation of synchrotron radiation from electrons accelerated in a petawatt-laser-generated plasma cavity. Phys Rev Lett 100:105006
62. Kneip S, McGuffey C, Martins JL et al (2010) Bright spatially coherent synchrotron X-rays from a table-top source. Nat Phys 6(12):980–983
63. Cipiccia S, Islam MR, Ersfeld B et al (2011) Gamma-rays from harmonically resonant betatron oscillations in a plasma wake. Nat Phys 7(11):867–871
64. Corde S, Phuoc KT, Fitour R et al (2011) Controlled betatron X-Ray radiation from tunable optically injected electrons. Phys Rev Lett 107:255003
65. Schnell M, Saevert A, Uschmann I et al (2013) Optical control of hard X-ray polarization by electron injection in a laser wakefield accelerator. Nat Commun 4
66. Albert F, Pollock BB, Shaw JL et al (2013) Angular dependence of betatron X-Ray spectra from a laser-wakefield accelerator. Phys Rev Lett 111:235004
67. Schwoerer H, Liesfeld B, Schlenvoigt HP et al (2006) Thomson-backscattered X Rays from laser-accelerated electrons. Phys Rev Lett 96:014802
68. Phuoc KT, Corde S, Thaury C et al (2012) All-optical compton gamma-ray source. Nat Photon 6(5):308–311
69. Chen S, Powers ND, Ghebregziabher I et al (2013) MeV-energy X rays from inverse compton scattering with laser-wakefield accelerated electrons. Phys Rev Lett 110:155003
70. Powers N, Ghebregziabher I, Golovin G et al (2014) Quasi-monoenergetic and tunable X-rays from a laser-driven Compton light source. Nat Photon 8:28–31
71. Pak A, Marsh KA, Martins SF et al (2010) Injection and trapping of tunnel-ionized electrons into laser-produced wakes. Phys Rev Lett 104:025003
72. McGuffey C, Thomas AGR, Schumaker W et al (2010) Ionization induced trapping in a laser wakefield accelerator. Phys Rev Lett 104:025004
73. Debus AD, Bussmann M, Schramm U et al (2010) Electron bunch length measurements from laser-accelerated electrons using single-shot THz time-domain interferometry. Phys Rev Lett 104:084802
74. Brunetti E, Shanks RP, Manahan GG et al (2010) Low emittance, high brilliance relativistic electron beams from a laser-plasma accelerator. Phys Rev Lett 105:215007
75. Lundh O, Lim J, Rechatin C et al (2011) Few femtosecond, few kiloampere electron bunch produced by a laser-plasma accelerator. Nat Phys 7(3):219–222
76. Gonsalves A, Nakamura K, Lin C et al (2011) Tunable laser plasma accelerator based on longitudinal density tailoring. Nat Phys 7(11):862–866
77. Buck A, Wenz J, Xu J et al (2013) Shock-front injector for high-quality laser-plasma acceleration. Phys Rev Lett 110:185006
78. Corde S, Thaury C, Lifschitz A et al (2013) Observation of longitudinal and transverse self-injections in laser-plasma accelerators. Nat Commun 4
79. Yan W, Chen L, Li D et al (2014) Concurrence of monoenergetic electron beams and bright X-rays from an evolving laser-plasma bubble. Proc Natl Acad Sci USA 111(16):5825–5830
80. Hua J (2014) Recent progress on plasma-based acceleration at Tsinghua university. AAC2014, San Jose CA, USA. Accessed 14–18 July 2014

Chapter 2
Phase Space Dynamics of Injected Electron Beams in Ionization Injection

2.1 Introduction

Plasma based wakefield acceleration has made great strides worldwide in the past decade [1]. In the laser driver case, energy gains up to 4 GeV and energy spreads of a few percent have been achieved using cm scale plasmas [2–10]; in the beam driver case, an energy gain more than 42 GeV has been demonstrated in a meter-scale plasma [11–13]. These have inspired great interest in plasma accelerators as drivers for compact coherent light sources for science and technology on the one hand and TeV-class colliders with much smaller footprint than the current RF-based accelerators for high energy physics on the other. However these applications require beams that have extremely small emittance and very low energy spread. Therefore the next challenge for plasma accelerators is to perfect methods for controlled injection of electron beams with very low emittance and low energy spread. Many novel methods for controlled injection have been proposed and demonstrated, such as injection by using plasma density ramps [14, 15], by colliding laser pulses [16–18] and by ionization induced injection [5, 19–30]. For each of these injection methods, there is an associated injection process that involves special 6D phase space evolution of the injected electron beam. It turns out that understanding the details of these phase space evolution is the first key step for perfecting the injection methods. In this chapter, we will present a systematical study of the phase space dynamics of the trapped electrons in ionization injection method.

Generally speaking, in ionization injection, electrons are produced inside the wake by tunneling ionization by the electric field of laser pulses or by the combined field of the beam-driver and the wake, where they can be more easily captured and accelerated. In 2006, the first ionization injection experiment was demonstrated at SLAC [19]. A 28.5 GeV, ultrashort electron beam is tightly focused and sent into an oven with lithium vapor and helium gas. The first electron of Li with a low ionization potential (IP, 5.39 eV) is tunneling ionized by the electric field of the front part of the beam driver and pushed out radially, forming the wake. The first electron of He with a relative high IP (24.6 eV) is released within the wake when it experiences the

© Springer Nature Singapore Pte Ltd. 2020
X. Xu, *Phase Space Dynamics in Plasma Based Wakefield Acceleration*,
Springer Theses, https://doi.org/10.1007/978-981-15-2381-6_2

peak of the tightly focused beam driver. These electrons from He could be trapped if the driver beam is intense enough. A trapping threshold of the wake amplitude is inferred from the experiments and confirmed by a 1D theoretical model and 2D PIC simulations. In 2009, ionization injection experiments driven by a single laser pulse have been demonstrated at UCLA [20] and University of Michigan [21]. In the UCLA experiment, an ultrashort 800 nm laser with a normalized vector potential $a_0 = 2$ is sent into a mixture of helium and nitrogen gases. The leading edge of the intense laser pulse is strong enough to ionize the two electrons of He and the first five L-shell electrons of N. These electrons are expelled by the ponderomotive force of the laser pulse, forming a nonlinear plasma wake. The two K-shell electrons of N^{5+} with much higher IPs (552 and 667 eV, respectively) remain bounded until meeting the peak intensity of the laser. After released inside the wake, the two K-shell electrons slip back to the tail of the wake and keep gaining energy. Trapping can occur if they obtain enough forward velocity to catch the wake before slipping out of the first bucket. The above process has been well reproduced in 3D PIC simulations [20]. Furthermore, a 3D injection threshold has also been derived in the quasi-static limit [20, 31]:

$$\delta\psi \equiv \psi_{inj} - \psi_{birth} \approx -1 \tag{2.1}$$

where $\psi \equiv e/mc^2(\phi - v_\phi/cA_z)$ is the normalized pseudo-potential of the wake, ϕ is the scalar potential, A_z is the axial vector potential and v_ϕ is the phase velocity of the wake. ψ_{birth} and ψ_{inj} represent the pseudo-potential where the electron is born and where the electron is injected ($v_z = v_\phi$), respectively.

The electrons are continuously ionized and injected if the trapping condition remains satisfied. If the total charge of the trapped electrons becomes substantial, the beam loading effect [32, 33] could change the wake structure and stop the trapping. The evolution of the driver, e.g., self-focusing or diffraction, could also change the wake structure and the ionization position and may stop the trapping [34, 35]. In typical scenarios of ionization injection, the injection distance, defined as the length where trapping occurs, is much larger than the plasma skin depth.

From the trapping threshold, one can see that the trapping process is mainly determined by the longitudinal dynamics. Besides this longitudinal motion, the trapped electrons also conduct transverse motions under the fields of the lasers and the transverse wakefield. First, the electrons are ionized by the strong electric field of the lasers with small initial momentum. Then, they respond to the electromagnetic field of the lasers with fast oscillations. After the laser pulses passing by, the electrons obtain certain amount of transverse momentum, which is defined as the residual momentum. The electrons conduct transverse betatron oscillations in the linear focusing field of the nonlinear wake [36]. During the whole injection process, the electrons are ionized and injected continuously, at the same time conduct the transverse motions, leading to a complicated phase space dynamics that determines the final quality of the trapped beams.

Since demonstrated experimentally, ionization injection has attracted much attention due to its simplicity and potential for high quality beam generation. Many vari-

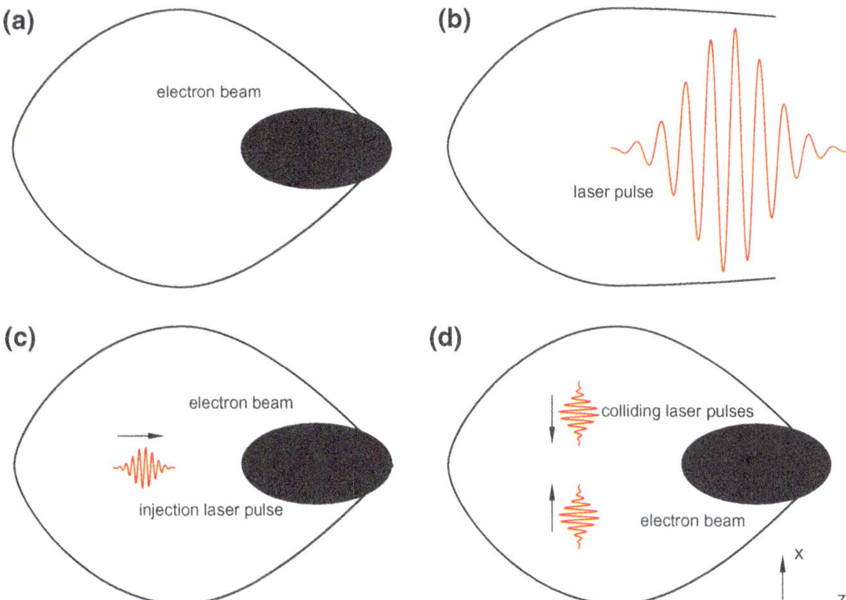

Fig. 2.1 Four scenarios of ionization injection. **a** A single beam pulse; **b** A single laser pulse; **c** Beam driver and laser longitudinal injection: an injection laser co-propagates with the wake; **b** Beam driver and laser transverse injection: two transversely propagating laser pulses collide in the wake

ants have been proposed and examined through simulations as shown in Fig. 2.1. Besides single pulse ionization injection [5, 19–21] (Fig. 2.1a, b), multiple pulses are utilized to separate the wake excitation and ionization processes [22–24, 26, 37](Fig. 2.1c, d). In a single laser pulse ionization injection scheme, the laser pulse is used to excite the wake and to release the trapped electrons simultaneously, as shown in Fig. 2.1b. This scheme has an obvious drawback for generating low emittance beams: to drive a large wake for trapping to occur [20], normalized laser vector potential a_0 greater than 1 is needed, which leads to a large residual momentum on the order of $\sim a_0$ for the trapped electrons along the laser polarization direction. At the same time, the radius of the ionized region of trapped electrons, which is connected with the radius of the driving laser, is on the order of a few laser wavelengths or a few μm. The transverse ponderomotive force for such a radius can also lead to residual momentum in both transverse directions. These residual momentum and initial radius lead to a normalized emittance on the order of 1 μm in the both transverse directions. To obtain a much smaller emittance through ionization injection, a possible strategy is to separate the wake excitation process and the injection process so that the injection laser can have a much lower vector potential and much smaller spot size. For example, an electron beam drives the wake and a laser pulse ionizes the injected electrons [22, 37]. Due to the much lower electric field of the electron beam driver, ionization of low IP atoms (e.g., H or Li) are used to excite the wake while the modest IP helium atoms (e.g., He) can be used to provide the injected electrons, which requires a much lower laser intensity. For example, the intensity threshold to ionize the first electron of He is $\sim 10^{15}$ W/cm^2 which corresponds to $a_0 \sim 0.02$ for an 800 nm wavelength laser. On the other hand for a single laser driver, the intensity threshold to ionize N^{5+} requires an intensity of $\sim 10^{19}$ W/cm^2 or $a_0 \sim 2$ for an 800 nm laser. Also, laser spot sizes much smaller than the transverse size of the wake can be used to reduce the radius of the ionized region. The use of lower a_0 and smaller spot sizes lead to much reduced normalized emittances of $\sim$30 nm [22] when using a single co-propagating laser pulse in a wake driven by a particle beam (Fig. 2.1c). Furthermore, if the injection distance can be limited through the use of two transversely colliding laser pulses then emittances less than 10 nm and slice energy spread as low as $\sim$10 keV can potentially be achieved [23] (Fig. 2.1d). These exciting results for high quality electron beam generation inspire people to study the ionization injection process at a much deeper level to answer the following questions: what physics determines the beam quality in ionization injection, such as the emittance and the energy spread? How does the emittance evolve in the whole injection process? Only a detailed understanding of the phase space dynamics of the trapped electrons can answer these critical questions, and this will be the main subject of this chapter.

In the following sections, a systematical study of the phase space dynamics in ionization injection will be presented. In Sect. 2.2 the nonlinear photoionization process is examined. In Sect. 2.3, the residual momentum of the trapped electrons is evaluated. In Sect. 2.4 a theoretical model based on single particle motion in the nonlinear wake is derived. In Sects. 2.5 and 2.6 two key phase mixing processes, namely the transverse and longitudinal phase mixing, are described respectively. In Sect. 2.7, the self space charge effects of the trapped electrons are studied. In Sect. 2.8, a high quality electron beam generation scheme using multi laser pulses with different wavelengths is studied. The intrinsic phase space discretization in laser triggered ionization injection is analyzed in Sect. 2.8.1 and a summary is provided in Sect. 2.10.

2.2 The Photoionization Process

Ionization of atoms and ions in intense laser fields has been studied extensively in the past few decades [38, 39]. In 1964, Keldysh showed that the nonlinear photoionization can be characterized by the Keldysh parameter $\gamma_K = \sqrt{I_p/2U_p}$, where I_p is the ionization potential, $U_p = E_0^2/4\omega_0^2$ is the average energy of electron oscillations in the laser field, E_0 is the amplitude of the laser field strength, and ω_0 is the laser frequency [40]. Note that atomic units are used in this section, i.e., energy is normalized to $\alpha^2 mc^2$, time to $r_0/\alpha c$, length to r_0, where α is the fine structure constant, m the mass of the electron, c the speed of light, and r_0 is the Bohr radius. Tunneling ionization takes place when $\gamma_K \ll 1$, where the ionization rate can be satisfactorily

Table 2.1 The Keldysh parameters and the momentum from tunneling in several typical ionization injection schemes. The lasers are linearly polarized

Parameters	Case A	Case B	Case C	Case D
Laser wavelength λ_0 (nm)	800	800	400	200
Normalized laser vector potential a_0	2.0	0.04	0.09	0.01
The atoms or ions	N^{5+}	He	O^{5+}	He
IP (eV)	552	24.6	138	24.6
γ_K	0.023	0.245	0.258	0.979
$\sigma_{p_\parallel,\text{tunnel}}$ (mc)	0.60	0.018	0.023	4.4×10^{-3}
$\sigma_{p_\perp,\text{tunnel}}$ (mc)	8.1×10^{-3}	2.5×10^{-3}	3.4×10^{-3}	2.5×10^{-3}

predicted by the Ammosov-Deloine-Krainov (ADK) model [41]. In the tunneling limit, the ionization rate is of a simple form with exponential accuracy

$$w \propto \exp\left[-\frac{2(2I_p)^{3/2}}{3E_0}\right] \tag{2.2}$$

While for $\gamma_K \gg 1$ multiphoton process dominates where several photons of energy below the ionization threshold may actually combine their energies to ionize an atom [40]. The ionization rate in the multiphoton limit depends on the field strength according to the power law

$$w \propto E_0^{2K} \tag{2.3}$$

where K is the number of absorbed photons.

In ionization injection schemes, the typical Keldysh parameter is smaller than 1, i.e., $\gamma_K \lesssim 1$, as cases A, B and C shown in Table 2.1. In cases where short wavelength laser is used, since $\gamma_K \propto \omega_0$, the Keldysh parameter could be around 1 as shown in case D in Table 2.1, which is in the regime of nonadiabatic tunneling.

Recently, a nonadiabatic model is obtained by Yudin and Ivanov [42] to give the ionization rate for arbitrary value of the Keldysh parameter γ_K within the usual strong-field approximation. The detailed derivations and the expressions of the ionization probability can be found in Ref. [42]. The ionization probability in half a laser cycle given by the nonadiabatic model and the ADK model are compared in Fig. 2.2. One can see the ADK model is a good approximation even when $\gamma_K \sim 1$.

Within one laser cycle, the ionization probability calculated from the two models reaches its maximum at the peak of the electric field. This phase-dependent ionization probability is the characteristic of the tunneling ionization. In the multiphoton regime with $\gamma_K \gg 1$, the ionization probability is phase-independent. The phase-dependent ionization probability in the tunneling regime could discrete the phase space of the

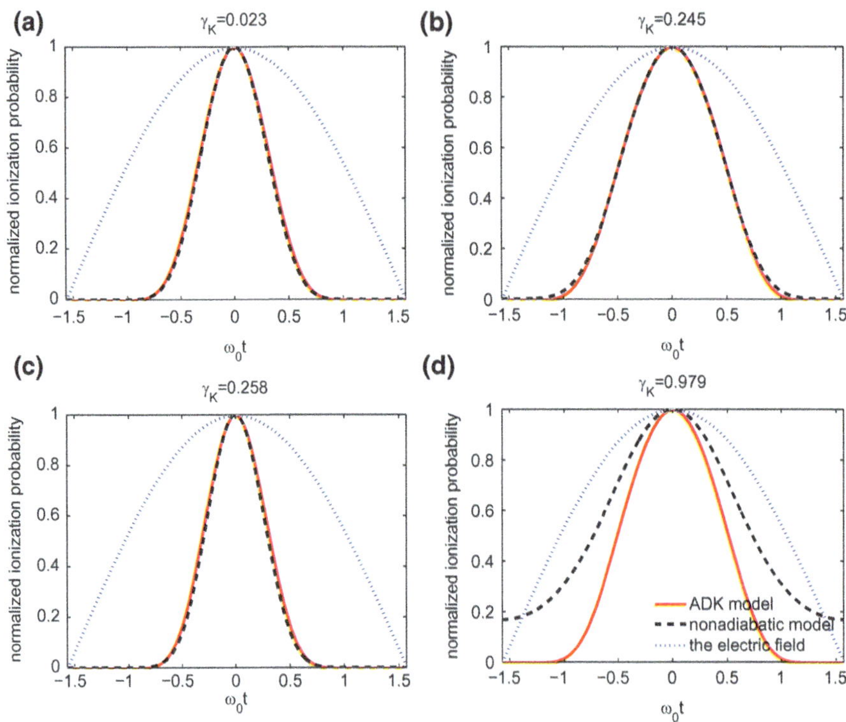

Fig. 2.2 Comparisons of the ionization probability between the ADK model and the nonadiabatic model. **a**, **b**, **c** and **d** are corresponding to the four cases listed in Table 2.1

injected electrons when they are freed and this discretization in the initial phase space can be mapped to the final phase space based on the trapping condition Eq. 2.1, leading to some interesting phase space structures. This intrinsic phase space discretization in laser triggered ionization injection will be discussed in Sect. 2.9.

2.3 The Residual Momentum

The residual momentum originates from two sources: one is the photoionization process and the other is the motion in the laser pulses. After released, the electrons obtain certain amount of momentum from the quantum ionization process and start to move under the combined fields of the lasers and the wake. In ionization injection, the laser pulses are used to trigger the ionization, and typically the field of the lasers is much larger than the wakefield in the region where the ionization occurs. The self space charge field of the freed electrons is also much lower than the field of the lasers (e.g., 0.16 TV/m corresponding to the electric field of a 800 nm laser with $a_0 = 0.04$).

Therefore, the motion of the ionized electron within the laser is dominated by the electromagnetic field of the laser pulses.

2.3.1 Initial Momentum from the Tunneling Ionization

The energy and momentum distribution of the electrons for the tunnel ionization of atoms by strong low-frequency radiation was obtained in Refs. [38, 43] with different approaches. For a linearly polarized laser, they showed the momentum distribution along and perpendicular to the field polarization direction are

$$w(p_\parallel, p_\perp) = w_0 \exp\left(-\frac{p_\parallel^2 \gamma_K^3}{3\omega_0} - \frac{p_\perp^2 \sqrt{2I_p}}{E_0}\right) \tag{2.4}$$

It is straightforward to obtain the rms value of $p_\parallel$ and $p_\perp$ as

$$\sigma_{p_\parallel,tunnel} = \frac{\int dp_\parallel p_\parallel^2 \int dp_\perp w(p_\parallel, p_\perp)}{\int dp_\parallel dp_\perp w(p_\parallel, p_\perp)} = \sqrt{\frac{3\omega_0}{2\gamma_K^3}} = \frac{1}{\omega_0}\left(\frac{9E_0^6}{32I_p^3}\right)^{1/4} \tag{2.5}$$

$$\sigma_{p_\perp,tunnel} = \frac{\int dp_\perp p_\perp^2 \int dp_\parallel w(p_\parallel, p_\perp)}{\int dp_\parallel dp_\perp w(p_\parallel, p_\perp)} = \left(\frac{E_0^2}{8I_p}\right)^{1/4} = \frac{\gamma_K}{\sqrt{3}}\sigma_{p_\parallel,tunnel} \tag{2.6}$$

For chosen atoms or ions, the momentum along the polarization direction is inversely proportional to the laser frequency, i.e, $\sigma_{p_\parallel,tunnel} \propto \omega_0^{-1}$, while the momentum perpendicular to the polarization direction is independent of the laser frequency. The rms values of the momentum in the tunneling ionization for several typical parameters are listed in Table 2.1. One can find for typical parameters the momentum along the polarization direction is smaller than the normalized vector potential of the laser pulse and the momentum perpendicular to the polarization direction is much smaller than a_0, i.e., $\sigma_{p_\parallel,tunnel} < a_0$, $\sigma_{p_\perp,tunnel} \ll a_0$.

2.3.2 The Momentum from the Lasers: Longitudinal Injection

After released, the electrons start to move in the electromagnetic field of one or more laser pulses. Certain amount of momentum is obtained by the electron after the laser pulses passing. The motion and the obtained momentum can be very different depending on the configuration of the laser pulses. In the longitudinal injection case (Fig. 2.1b, c), only one laser pulse is used to free the electrons. The motion of an electron in a single laser pulse has been studied thoroughly [44]. We will give a simple review of the residual momentum in this case. In the transverse injection case (Fig. 2.1d), two colliding lasers are used and the electron motion is described by a non-integrable Hamiltonian [45]. The electron motion in the two colliding lasers is complicated depending on the intensity and the duration of the lasers. If the intensity of the lasers is low, the conclusions from a single laser case are still approximately valid. The residual momentum in the two transverse directions are much lower than $a_0 mc$ because the lasers are polarized in the wake propagation direction. If the intensity is sufficiently high, the standing wave would impart a large momentum ($\sim a_0 mc$) in the propagation direction of the lasers. If the intensity is higher than a certain criteria, stochastic heating would play an important role in the electron motion [45–48]. Note that only the residual momentum in the two transverse directions are concerned.

In a single laser pulse case, the residual momentum comes from two physical effects: the 1D residual drift and the ponderomotive acceleration [49]. We study the 1D residual drift first. The Hamiltonian for an electron in the electromagnetic field is

$$H = c\sqrt{m^2 c^2 + \left(\mathbf{P} - \frac{e}{c}\mathbf{A}\right)^2} + e\phi \tag{2.7}$$

Assuming a 1D plane wave as $\mathbf{A}(t, z) = \hat{x}(mc^2/e)a_0 f(t - z/v_\phi)\sin(\omega_0 t - k_0 z)$, where $f(t)$ is the normalized envelop, and $v_\phi = \omega_0/k_0$ is the phase velocity of the plane wave. It is straightforward to see there are three conservation quantities of the motion, i.e., the two transverse canonical momentum $P_{x,y} \equiv p_{x,y} + (e/c)A_{x,y}$ and a combination of the energy and the longitudinal momentum as $\gamma mc - v_\phi/cp_z$. Therefore, the residual momentum after the laser passes can be deduced from the conservation of the transverse canonical momentum as $p_{x,y} = p_{x0,y0} + (e/c)A_{x0,y0}$, where $p_{x0,y0}$ is the momentum from the ionization and $A_{x0,y0}$ is the initial vector potential when the electron is born. This 1D residual drift is only strictly held in a 1D plane wave [50]. Even for finite width waves this gives a reasonable estimate for typical laser spot sizes. Due to the fact that most electrons are released near the peak of the laser electric field by tunneling (as shown in Fig. 2.2), where the vector potential is close to zero, this residual drift points in the polarization direction and it is therefore smaller than the simple estimation based on the peak value of the vector potential a_0 [51]. It can be also shown that for a linearly polarized laser polarized along $\hat{x}$, the electric field scales as ϵa_0 in the $\hat{z}$ direction (the propagation direction) and as $\epsilon^2 a_0$ in the $\hat{y}$ direction where $\epsilon = 1/k_0 w_0$ [44]. Therefore, for a linearly polarized laser the residual drift from this effect is very asymmetric between the two transverse directions.

For low emittance electron beam generation, the laser pulse is focused tightly to restrict the ionization volume. Therefore, high-dimensional effects play an important role in the electron motion. The contribution of the ponderomotive force to the residual momentum can be estimated as $p_{x,y} \sim (a_0^2 c\tau_{FWHM})/(\bar{\gamma}w_0)mc$, where $\bar{\gamma}$ is the average relativistic factor of the electron when it moves in the laser, τ_{FWHM} is the full-width-half-maximum (FWHM) duration of the laser pulse intensity and w_0 is the spot size of the laser. For relativistically intense pulse $a_0 \gtrsim 1$ or cigar shaped pulses where $a_0^2 c\tau_{FWHM} \gg w_0$ the ponderomotive acceleration may be the dominant cause for the residual drift. One can see this contribution is equal in the two transverse directions.

2.3.3 The Momentum from the Lasers: Transverse Injection

In the transverse injection schemes, the intensity of the lasers is optimized to be as low as possible for low emittance electron beam generation [23], typically $a_{1,2} < 0.1$, where $a_{1,2}$ are the normalized vector potentials of the two lasers. In most cases, the intensity of the lasers is much lower than the criterion where stochastic heating can play a role. For example, the criterion for local stochastic motion have been estimated to be about $a_1 a_2 = 1/16$ by Mendonca [45]. Therefore, in the following content, we will not consider the stochastic motion.

For simplicity, two counter-propagating plane wave with same longitudinal envelope and equal intensity are assumed. The vector potential is $\mathbf{A}(t, x) = \hat{z}(mc^2/e)a_0$ $[f(t - x/c)\sin(\omega_0 t - k_0 x) + f(t + x/c)\sin(\omega_0 t + k_0 x)]$, where $f(t)$ is the normalized envelope. It is straightforward to see the conservation of the canonical momentum in the polarized direction, i.e., $P_z \equiv p_z + (e/c)A_z = \text{const}$. Neglecting the momentum from the ionization process, we can get $p_z = p_{z0} + (e/c)A_{z0} - (e/c)A_z \approx -(e/c)(A_z - A_{z0})$. The motion equation in the $\hat{x}$-direction is

$$\frac{\mathrm{d}p_x}{\mathrm{d}t} = \frac{\mathrm{d}P_x}{\mathrm{d}t} = -\frac{\partial H}{\partial x} = \frac{ep_z}{\gamma mc}\frac{\partial A_z}{\partial x} \tag{2.8}$$

Substitute the p_z expression, we can get

$$\frac{\mathrm{d}p_x}{\mathrm{d}t} + \frac{1}{2}\frac{e^2}{\gamma mc^2}\frac{\partial}{\partial x}(A_z - A_{z0})^2 = 0 \tag{2.9}$$

The typical motion of an electron is shown in Fig. 2.3. The momentum in the polarized direction oscillates quickly due to the oscillation of the vector potential the electron experiences. In the propagation direction of the lasers, the electron also conducts oscillation with a long period. After the lasers pass it, the electron can obtain a momentum which is on the order of the oscillation amplitude in the laser propagation direction.

In the transverse injection schemes, in order to generate electron beam with low emittance the intensity of the laser needs to be as small as possible, i.e., $a_0 \ll 1$. Therefore the electron motion in the lasers is non-relativistic, then Eq. 2.9 can be simplified as

$$\frac{\mathrm{d}^2 x}{\mathrm{d}t^2} + \frac{1}{2}\frac{e^2}{m^2 c^2}\frac{\partial}{\partial x}(A_z - A_{z0})^2 = 0 \tag{2.10}$$

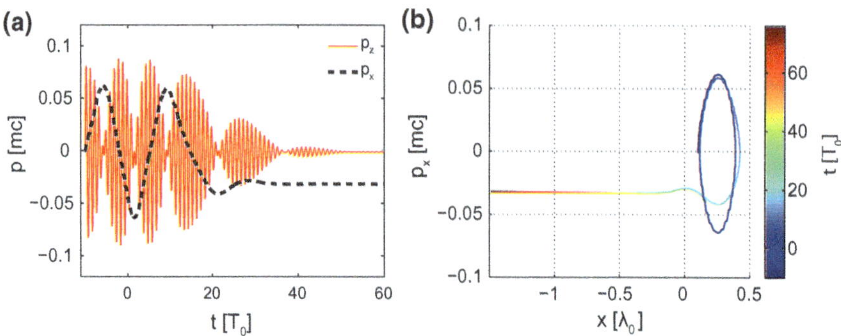

Fig. 2.3 The trajectory of a test particle. **a** p_x, p_z versus t; **b** p_x versus x. The color represents the time. Two laser pulses with gaussian envelopes are used, i.e., $f(t) = \exp\left[-t^2/(2\tau^2)\right]$. The parameters are: $a_0 = 0.06$, $\tau = 17T_0$, $x_0 = 0.1\lambda_0$, $t_0 = -10T_0$, where T_0 is the laser period

Due to the fact that most electrons are released near the peak of the laser electric field by tunneling (as shown in Fig. 2.2), where the vector potential is close to zero, $A_{z0} = 0$ is a good approximation when studying the electron motion. Substitute the expression of the vector potential into Eq. 2.10 and neglect the high frequency component of t, the motion in x-direction can be simplified as

$$\frac{\mathrm{d}^2 x}{\mathrm{d}t^2} - k_0 c^2 a_0^2 f(t - x/c) f(t + x/c)\sin 2k_0 x = 0 \tag{2.11}$$

where we assume the characteristic time of the envelop τ is much longer than the laser period, i.e., $\omega_0\tau \gg 1$. One can see the electron conducts pendulum motion with a slowly varying period. The equilibrium position of the pendulum motion is $x_{eq} = \lambda_0(1 + 2n)/4$, where $n = 0, \pm 1, \pm 2, \ldots$. As we know, the electrons are mostly released around $x_0 = n/2\,\lambda_0$, therefore they oscillate around the nearest equilibrium position with an amplitude close to $\lambda_0/4$ and the amplitude of the oscillation momentum is close to $a_0 mc$. If the pulse duration is short, the lasers pass it before the electron completes 1/4 period, therefore the residual momentum p_x is much lower than $a_0 mc$; if the pulse duration is long enough, the electron would obtain a large residual momentum which is on the order of $a_0 mc$ as shown in Fig. 2.3. Therefore, to reduce the residual momentum in $\hat{x}$ direction, the duration of the lasers must be shorter than some threshold.

If we neglect the field envelope, the period of the pendulum equation Eq. 2.11 can be expressed as $T(x_0) = 2\sqrt{2}K\left(\sin k_0 x_0\right)/(\omega_0 a_0)$, where x_0 is the initial position when the electron is released, and $K(k)$ is the complete elliptic integral of the first kind defined by $K(k) = F(\pi/2, k) = \int_0^{\pi/2}(1 - k^2\sin^2 u)^{-1/2}du$. From numerical calculation, we know $T(\lambda_0/4) \approx 1.2T(0)$, where $T(0) = \sqrt{2}\pi/(\omega_0 a_0)$ is the period under the small-angle approximation. Therefore, we can define the pulse length threshold as

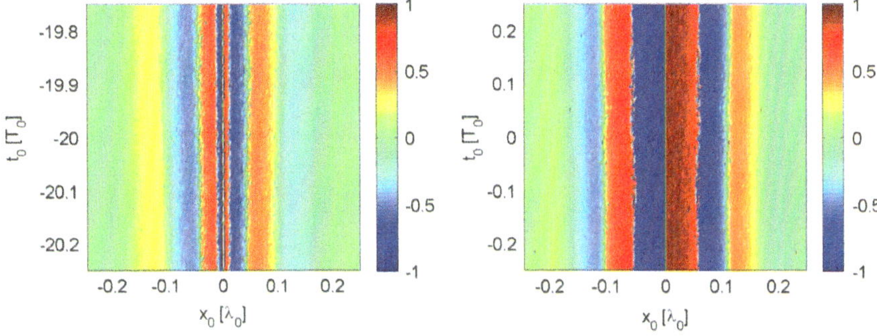

Fig. 2.4 The dependence of the residual p_x on the birth position x_0 and the birth time t_0. Lasers with gaussian envelopes are used. The laser pulses parameters: $a_0 = 0.06$, $\tau = 17T_0 \gg \tau_{crit} \approx 3.4T_0$

$$\tau_{crit} = \frac{T(\lambda_0/4)}{4} \approx \frac{1.3}{\omega_0 a_0} \tag{2.12}$$

In a convenient form, the threshold is

$$\tau_{crit}[\text{fs}] \equiv \frac{2.3}{E_0[\text{TV/m}]} \tag{2.13}$$

where E_0 is the peak electric field of the laser pulses. One can see the criterion is determined by the field strength. Therefore, for chosen atoms or ions, the field strength for ionization is fixed, leading to a fixed duration criterion whatever the laser frequency is.

When the pulse duration is less than the threshold, i.e., $\tau < \tau_{crit}$, the residual momentum is lower than $a_0 mc$. When $\tau \gg \tau_{crit}$, the dependence of the residual momentum p_x over $a_0 mc$ on the birth position and the birth time of the electron is shown in Fig. 2.4. One can see the residual momentum has strong dependence on the birth position and weak dependence on the birth time. Base on the ionization probability calculation, the electrons are mostly released near the maximum electric field, i.e., $t_0 = m/2T_0$, $x_0 = n/2\lambda_0$, where m, n are integers. These ionized electrons would obtain a residual momentum on the order of $a_0 mc$ in the laser propagation direction.

We also conduct 1D PIC simulations to confirm the above analytical analysis. The root-mean-squared (rms) value of the residual momentum under different conditions are shown in Fig. 2.5. While $\tau < \tau_{crit}$, the rms value of the residual momentum increases linearly with the pulse length; while $\tau \geq \tau_{crit}$, the rms value of the residual momentum saturates around $\sim 0.7 a_0 mc$.

From the above analysis, we know the residual momentum in the propagation direction in two colliding lasers configuration is not simply $\epsilon a_0 mc$ even in 1D. The residual momentum in the propagation direction can be as large as $a_0 mc$ depending on the intensity and duration of the lasers.

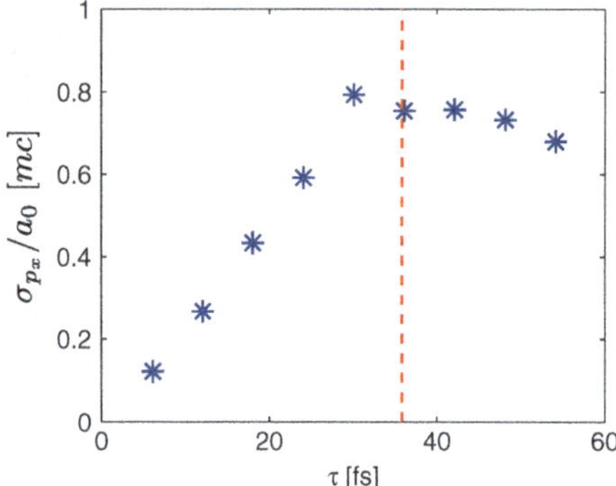

Fig. 2.5 1D OSIRIS simulations: The rms residual momentum in the laser propagation direction while scanning the laser pulse duration. The red line shows the critical duration calculating from Eq. 2.13. The parameters of the longitudinal gaussian lasers: $\lambda_0 = 800$ nm, $a_0 = 0.016$. And helium is initialized to be ionized

2.3.4　The Thermal Emittance

In traditional accelerators, after flying away from the cathode, the photo electrons obtain certain amount of transverse momentum due to various physical effects, e.g., electron-electron scattering, the Schottky effect, surface roughness of the cathode [52]. The emittance immediately after the electrons are emitted is defined as the thermal emittance which is the minimum value of the emittance in the whole beam line. Similarly, we define the thermal emittance in ionization injection as the emittance after the injection lasers passed the electrons. In order to remove the effect of different ionization time, we calculate the thermal emittance by assuming all electrons are freed at the same time (Fig. 2.6).

Due to the small transverse momentum and the short interaction time between the electrons and the lasers, the distribution in the transverse real space can be assumed to be unchanged when the lasers pass the electrons. Therefore, the thermal emittance ϵ_{th} can be approximated as the product of the transverse size σ_{x0} when the electrons are born and the residual momentum $\sigma_{p_{x,th}}$, i.e., $\epsilon_{th} \approx \sigma_{x0}\sigma_{p_{x,th}}$. In the following sections, we will show the final emittance of the injected electrons can be much larger than the thermal emittance due to the complicated phase space dynamics.

2.4　Single Particle Motion in the Nonlinear Wake

After the lasers passing by, the electron begins to move in the wakefield. The nonlinear theory in the blowout regime showed the force on an electron in the ion column can be written as [53, 54]

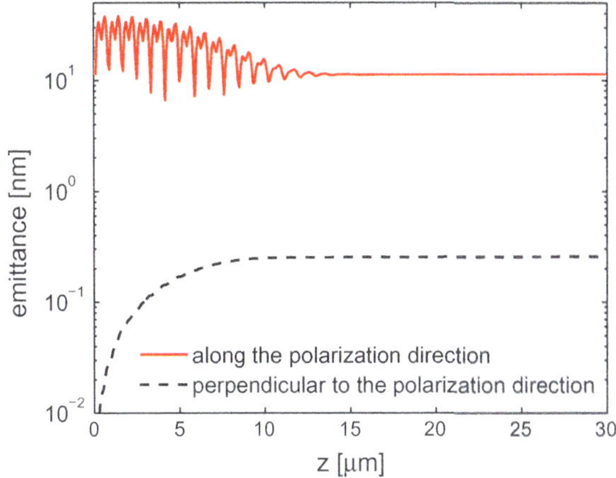

Fig. 2.6 2D OSIRIS simulations: the thermal emittance in the longitudinal injection case. The laser parameters: $\lambda = 800$ nm, $a_0 = 0.04$, $w_0 = 3\,\mu$m, $\tau \approx 30$ fs. And helium is initialized to supply the electrons. The thermal emittance in the transverse directions are: $\epsilon_{th,\parallel} = 11$ nm, $\epsilon_{th,\perp} = 0.3$ nm. The initial spot size and the residual momentum are $\sigma_{x0} = 1.4\,\mu$m, $\sigma_{p_{th,\parallel}} = 0.008$, $\sigma_{p_{th,\perp}} = 2.6 \times 10^{-4}$. Note that the electrons are assumed to be ionized at rest in OSIRIS

$$F_\perp = -\left[\frac{1}{2} - \frac{(1 - v_z)}{2}\frac{d^2}{d\xi^2}\psi_0(\xi) - (1 - v_z)\lambda(\xi)\right] r \tag{2.14}$$

$$F_z = -\frac{d}{d\xi}\psi_0(\xi) \tag{2.15}$$

where $\lambda(\xi) = \int_0^{r \gg \sigma_r} r\,dr\,n_b$ represents the current of the beam driver which is close to zero in the region where the ionization occurs and $\psi_0(\xi)$ is the pseudo-potential on axis. Note that in the following content we adopt the normalized units, where time is normalized to ω_p^{-1}, length to c/ω_p, velocity to c, mass to the electron mass, m, and charge to electron charge, e, fields to $mc\omega_p/e$, potentials to mc^2/e, where ω_p is the plasma frequency. And new variables $\xi \equiv t - z$ and $s \equiv z$ are introduced to replace the variables z and t for convenience.

The motion equation of an electron in a nonlinear wake is

$$\gamma\frac{d^2x}{dt^2} + \frac{d\gamma}{dt}\frac{dx}{dt} = F_x \tag{2.16}$$

After released, the forward velocity of the electron is accelerated to near the speed of light in several plasma skin depth, and then the longitudinal position is nearly locked in the wake, therefore some approximations can be applied, i.e., $d/dt \approx d/ds$, $\gamma \approx \gamma_0 + E_z s$ and $F_\perp \approx -gr$, where the coefficient $1/2$ is replaced by a parameter g for cases where the plasma electrons are not fully blown out in the wake. We also assume E_z is constant for each electron and g is constant for all electrons. Then, the motion equation can be rewritten as

$$\frac{d^2 x}{ds^2} + \frac{E_z}{\gamma_0 + E_z z}\frac{dx}{ds} + k_\beta^2 x = 0 \tag{2.17}$$

where $k_\beta = \sqrt{g/\gamma}$ is the betatron wave number.

The solution of Eq. 2.17 is

$$x = c_1 J_0\left(\sqrt{\frac{4g\gamma}{E_z^2}}\right) + c_2 Y_0\left(\sqrt{\frac{4g\gamma}{E_z^2}}\right) \tag{2.18}$$

$$\frac{dx}{ds} = -\sqrt{\frac{g}{\gamma}}\left[c_1 J_1\left(\sqrt{\frac{4g\gamma}{E_z^2}}\right) + c_2 Y_1\left(\sqrt{\frac{4g\gamma}{E_z^2}}\right)\right] \tag{2.19}$$

where the coefficients c_1, c_2 are determined by the initial conditions.

Assuming $p_x \approx \gamma dx/ds$, then the transformation between the positions in the transverse phase space can be written as,

$$\begin{pmatrix} x \\ p_x \end{pmatrix} = \begin{pmatrix} J_0\left(\sqrt{\frac{4g\gamma}{E_z^2}}\right) & Y_0\left(\sqrt{\frac{4g\gamma}{E_z^2}}\right) \\ -\sqrt{g\gamma}J_1\left(\sqrt{\frac{4g\gamma}{E_z^2}}\right) & -\sqrt{g\gamma}Y_1\left(\sqrt{\frac{2\gamma}{E_z^2}}\right) \end{pmatrix} \begin{pmatrix} J_0\left(\sqrt{\frac{4g\gamma_0}{E_z^2}}\right) & Y_0\left(\sqrt{\frac{4g\gamma_0}{E_z^2}}\right) \\ -\sqrt{g\gamma_0}J_1\left(\sqrt{\frac{4g\gamma_0}{E_z^2}}\right) & -\sqrt{g\gamma_0}Y_1\left(\sqrt{\frac{4g\gamma_0}{E_z^2}}\right) \end{pmatrix}^{-1} \begin{pmatrix} x_0 \\ p_{x0} \end{pmatrix}$$

$$= J(E_z, s)\begin{pmatrix} x_0 \\ p_{x0} \end{pmatrix} \tag{2.20}$$

where $J(E_z, s)$ is the Jacobian matrix of above transformation. It is straightforward to show that the Jacobian determinant of $J(E_z, s)$ is unity.

When $z \gg \left|\alpha^2 - \frac{1}{4}\right|$, the asymptotic expressions of the Bessel functions are

$$J_\alpha(z) \approx \sqrt{\frac{2}{\pi z}}\cos\left(z - \frac{\alpha\pi}{2} - \frac{\pi}{4}\right)$$

$$Y_\alpha(z) \approx \sqrt{\frac{2}{\pi z}}\sin\left(z - \frac{\alpha\pi}{2} - \frac{\pi}{4}\right)$$

In the following derivation, we use the asymptotic forms of Bessel functions, i.e.,

$$J(E_z, s) \approx \begin{pmatrix} \left(\frac{\gamma_0}{\gamma}\right)^{1/4}\cos\phi & \left(\frac{1}{g^2\gamma\gamma_0}\right)^{1/4}\sin\phi \\ \left(g^2\gamma\gamma_0\right)^{1/4}\sin\phi & \left(\frac{\gamma}{\gamma_0}\right)^{1/4}\cos\phi \end{pmatrix} \tag{2.21}$$

where $\phi = 2\sqrt{g}(\sqrt{\gamma} - \sqrt{\gamma_0})/E_z$ is the betatron phase of a particle starting at rest. One can note the Jacobian determinant of this approximated $J(E_z, s)$ is also unity.

To check the accuracy of the approximated solutions, we compare the numerical calculations of the motion equation Eq. 2.16 with the approximation expressions Eq. 2.21 as shown in Fig. 2.7. We can see there is a finite phase difference ($\sim \pi/4$) between the asymptotic solutions and the numerical calculations, but the amplitudes of the oscillations are similar.

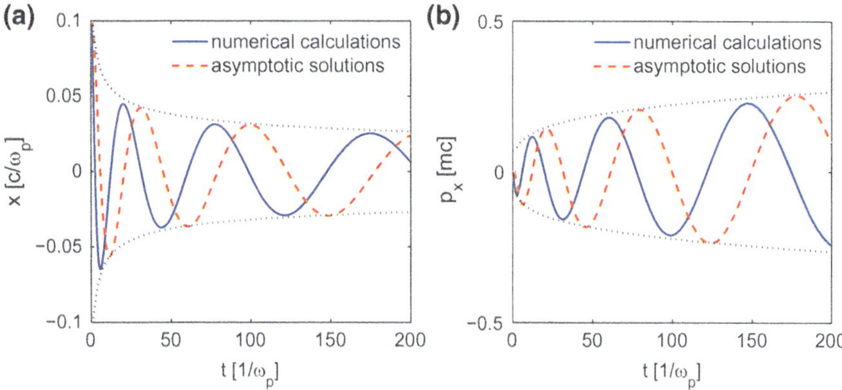

Fig. 2.7 The comparison between the numerical calculations of Eq. 2.16 and the analytical results Eq. 2.21. The black dotted lines show the analytical envelope of the betatron oscillation. The forces are: $F_\perp = -r/2$, $F_z = \xi/2$. The initial condition of the test particle is: $x = 0.1$, $p_x = 0$, $y = 0$, $p_y = 0$, $z = 0$, $p_z = 0$

2.5 Transverse Phase Mixing

In ionization injection, the electrons are freed and injected continuously within the injection distance. Due to the relatively long injection distance (on the order of tens of plasma skin depth), the first ionized electron has rotated one or more betatron periods when the last electron is ionized. Therefore, the electrons ionized at different times will have different betatron phases and they mix together in the transverse phase space. The normalized emittance, i.e., the transverse phase space area occupied by the electrons, can exhibit complicated evolution due to this transverse phase mixing. We find that three distinct stages exist in this evolution and that each stage can impact the final beam quality. In a typical case where the injection time is limited to few inverse plasma periods $(2\pi\omega_p^{-1})$ and the charge is low, these three stages are as follows. First, when ionization is occurring, the emittance of the injected beam grows quickly in time from the initial thermal emittance. Second, immediately following ionization, the emittance slowly decreases to a minimum value. Finally, the emittance again gradually increases to a saturated value. If the ionization time is more than $\sim\pi\omega_p^{-1}$, the emittance grows to the saturated level during the first stage including an oscillatory behavior before it slowly decreases.

This complicated evolution can be reproduced by solving the single particle motion equation. We integrate Eq. 2.16 for many test electrons. The acceleration gradient of each electron is randomly chosen from 0.9 to 1.1 and kept as constant over the whole time. In Fig. 2.8a we show electrons released at different times. After a group of electrons is released they begin to rotate in the transverse phase space. The first group (red) has a betatron phase ϕ_M, and the most recent group (purple) has a phase ϕ_m. Obviously, the area in phase space increases during the injection process due to each group of electrons having a different betatron phase.

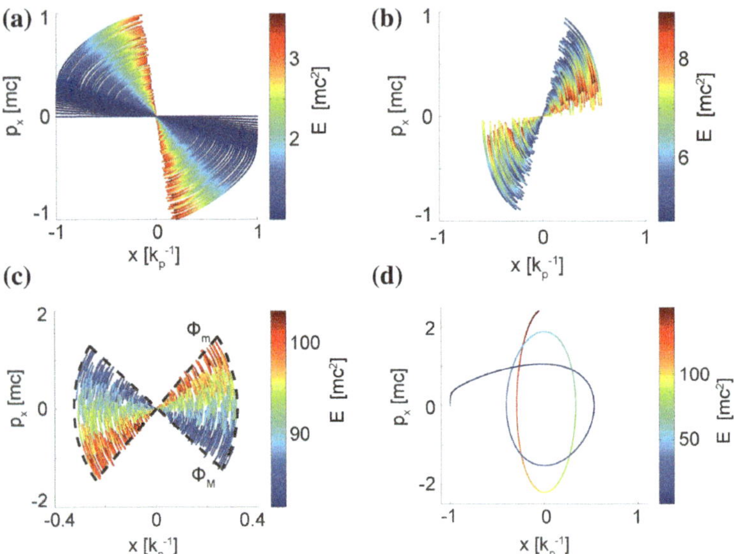

Fig. 2.8 Phase space snapshots from single particle model illustrating the transverse phase mixing process. **a–c** $x - p_x$ phase space at different times: **a** $t = 3$ when the ionization is just terminated, **b** $t = 8$, **c** $t = 94$. **d** $x - p_x$ phase space trajectory for a typical particle, where the color represents the relative factor γ. Note that at early time, γ is correlated with the ionization time. For simplicity, the electrons are born with finite transverse position and zero momentum (Reprinted from Ref. [55]. Copyright with kind permission from American Physical Society)

In Fig. 2.8b we show the phase space at a 'time' after the ionization has ceased. For simplicity, we consider a case where the injection time $\Delta s < \pi$. However, electrons at ϕ_M have a higher energy (due to being accelerated for a longer time) and hence lower betatron frequency than those electrons at ϕ_m. As a result, $\phi_M - \phi_m \equiv \Delta\phi$ gradually decreases and hence the emittance gradually decreases. Later in time, due to any spread in E_z, the electrons ionized at the same time develop a spread in phase. Eventually, the electrons at ϕ_M (ϕ_m) are those injected first (last) but which have experienced the smallest (largest) E_z. In this case, electrons at ϕ_M now rotate faster than those at ϕ_m, causing the emittance to gradually increase as shown in Fig. 2.8c.

In this section, we try to formulate this phase space dynamics. The electrons can be considered as many bunches which are ionized at different time s_{birth} and accelerated with different acceleration gradient. Assuming the initial phase space distribution of these bunches when they are released is the same as $f_0(x_0, p_{x0})$. And $f_E(E_z)$ is introduced to represent the distribution of the acceleration gradient. The distribution of the acceleration distance of these bunches can be described by a function $f_s(\hat{s})$, where $\hat{s} \equiv s - s_{birth}$. We assume the three distribution functions $f_0(x_0, p_{x0})$, $f_E(E_z)$ and $f_s(\hat{s})$ are independent with each other.

Based on the transformation in phase space positions Eq. 2.20, the emittance evolution can be obtained. First, by integrating over many bunches with different acceleration distances and different acceleration gradients, the covariance matrix of the transverse phase space coordinates (x, p_x) can be obtained as

$$
\begin{pmatrix} \sigma_x^2 & \sigma_{xp_x} \\ \sigma_{xp_x} & \sigma_{p_x}^2 \end{pmatrix} = \int d\hat{s}\, f_s(\hat{s}) \int dE_z\, f_E(E_z) \int\int dx\, dp_x\, f(x(E_z,\hat{s}), p_x(E_z,\hat{s}))
$$

$$
\begin{pmatrix} \langle (x - \langle x \rangle)^2 \rangle & \langle (x - \langle x \rangle)(p_x - \langle p_x \rangle) \rangle \\ \langle (x - \langle x \rangle)(p_x - \langle p_x \rangle) \rangle & \langle (p_x - \langle p_x \rangle)^2 \rangle \end{pmatrix}
$$

$$
= \int d\hat{s}\, f_s(\hat{s}) \int dE_z\, f_E(E_z) \int\int dx_0\, dp_{x0}\, f_0(x_0, p_{x0}) J(E_z, \hat{s})
$$

$$
\begin{pmatrix} \langle (x_0 - \langle x_0 \rangle)^2 \rangle & \langle (x_0 - \langle x_0 \rangle)(p_{x0} - \langle p_{x0} \rangle) \rangle \\ \langle (x_0 - \langle x_0 \rangle)(p_{x0} - \langle p_{x0} \rangle) \rangle & \langle (p_{x0} - \langle p_{x0} \rangle)^2 \rangle \end{pmatrix} J^T(E_z, \hat{s})
$$

$$
= \int d\hat{s}\, f_s(\hat{s}) \int dE_z\, f_E(E_z) \begin{pmatrix} J_{11}^2 \sigma_{x0}^2 + J_{12}^2 \sigma_{p_{x0}}^2 & J_{11}J_{21}\sigma_{x0}^2 + J_{12}J_{22}\sigma_{p_{x0}}^2 \\ J_{11}J_{21}\sigma_{x0}^2 + J_{12}J_{22}\sigma_{p_{x0}}^2 & J_{21}^2 \sigma_{x0}^2 + J_{22}^2 \sigma_{p_{x0}}^2 \end{pmatrix}
$$

$$\tag{2.22}$$

where $f(x(E_z,\hat{s}), p_x(E_z,\hat{s}))dx\,dp_x = f_0(x_0, p_{x0})dx_0\,dp_{x0}$ is used, and $\sigma_{x_0 p_{x0}} = 0$ is assumed.

Substitute the expressions of $J(E_z, \hat{s})$ and the distribution functions $f_0(x_0, p_{x0})$, $f_E(E_z)$ and $f_s(\hat{s})$, the evolution of the emittance which is $\epsilon_n = \sqrt{\sigma_x^2 \sigma_{p_x}^2 - \sigma_{xp_x}^2}$ can be integrated numerically. In the following content, we discuss the analytical results of the emittance evolution with reasonable approximations.

2.5.1 Emittance Evolution: Growth and Oscillation in the Injection Stage

In the initial stage, the spread of E_z plays a negligible role in the evolution of emittance, the acceleration gradient distribution can be approximated as

$$
f_E(E_z) \approx \delta(\bar{E}_z) \tag{2.23}
$$

where $\bar{E}_z$ is the average value of the acceleration gradient the electrons feel. Then the emittance can be obtained as,

$$
\epsilon_n^2 = \sigma_x^2 \sigma_{p_x}^2 - \sigma_{xp_x}^2
$$

$$
\approx \sigma_{x0}^4 \left[\int d\hat{s}\, f_s(\hat{s}) J_{11}^2 \int d\hat{s}\, f_s(\hat{s}) J_{21}^2 - \left(\int d\hat{s}\, f_s(\hat{s}) J_{11} J_{21} \right)^2 \right]
$$

$$
+ \sigma_{p_{x0}}^4 \left[\int d\hat{s}\, f_s(\hat{s}) J_{12}^2 \int d\hat{s}\, f_s(\hat{s}) J_{22}^2 - \left(\int d\hat{s}\, f_s(\hat{s}) J_{12} J_{22} \right)^2 \right]
$$

$$
+ \sigma_{x0}^2 \sigma_{p_{x0}}^2 \left[\int d\hat{s}\, f_s(\hat{s}) J_{11}^2 \int d\hat{s}\, f_s(\hat{s}) J_{22}^2 + \int d\hat{s}\, f_s(\hat{s}) J_{12}^2 \int d\hat{s}\, f_s(\hat{s}) J_{21}^2 \right.
$$

$$
\left. - 2 \left(\int d\hat{s}\, f_s(\hat{s}) J_{11} J_{21} \right) \left(\int d\hat{s}\, f_s(\hat{s}) J_{12} J_{22} \right) \right] \tag{2.24}
$$

To obtain the expression of the emittance, a rectangle distribution of $f_s(\hat{s})$ is assumed, then,

$$\int_{\hat{s}_i}^{\hat{s}_f} d\hat{s}\, f_s(\hat{s}) J_{11}^2 = \frac{1}{\hat{s}_f - \hat{s}_i} \left(\frac{\sqrt{\gamma\gamma_0}}{\bar{E}_z} + \frac{1}{4}\sqrt{\frac{\gamma_0}{g}}\sin2\phi \right)\Bigg|_{\hat{s}_i}^{\hat{s}_f}$$

$$\int_{\hat{s}_i}^{\hat{s}_f} d\hat{s}\, f_s(\hat{s}) J_{12}^2 = \frac{1}{\hat{s}_f - \hat{s}_i} \left(\frac{1}{g\bar{E}_z}\sqrt{\frac{\gamma}{\gamma_0}} - \frac{1}{4g^{3/2}\sqrt{\gamma_0}}\sin2\phi \right)\Bigg|_{\hat{s}_i}^{\hat{s}_f}$$

$$\int_{\hat{s}_i}^{\hat{s}_f} d\hat{s}\, f_s(\hat{s}) J_{11} J_{21} = \frac{1}{16(\hat{s}_f - \hat{s}_i)}\sqrt{\frac{\gamma_0}{g}}\left(4\sqrt{g\gamma}\cos2\phi - \bar{E}_z\sin2\phi \right)\Bigg|_{\hat{s}_i}^{\hat{s}_f}$$

$$\int_{\hat{s}_i}^{\hat{s}_f} d\hat{s}\, f_s(\hat{s}) J_{12} J_{22} = -\frac{1}{16g(\hat{s}_f - \hat{s}_i)}\sqrt{\frac{\gamma_0}{g}}\left(4\sqrt{g\gamma}\cos2\phi - \bar{E}_z\sin2\phi \right)\Bigg|_{\hat{s}_i}^{\hat{s}_f}$$

$$\int_{\hat{s}_i}^{\hat{s}_f} d\hat{s}\, f_s(\hat{s}) J_{21}^2 = \frac{\sqrt{\gamma_0}}{96\bar{E}_z(\hat{s}_f - \hat{s}_i)}\left[32g\gamma^{3/2} - 12\bar{E}_z^2\sqrt{\gamma}\cos2\phi + 3\bar{E}_z g^{-1/2}\left(\bar{E}_z^2 - 8g\gamma \right)\sin2\phi \right]\Bigg|_{\hat{s}_i}^{\hat{s}_f}$$

$$\int_{\hat{s}_i}^{\hat{s}_f} d\hat{s}\, f_s(\hat{s}) J_{22}^2 = \frac{1}{96\bar{E}_z\sqrt{\gamma_0}(\hat{s}_f - \hat{s}_i)}\left[32\gamma^{3/2} + 12\bar{E}_z^2 g^{-1}\sqrt{\gamma}\cos2\phi - 3\bar{E}_z g^{-3/2}\left(\bar{E}_z^2 - 8g\gamma \right)\sin2\phi \right]\Bigg|_{\hat{s}_i}^{\hat{s}_f}$$

The emittance can be found by substituting the above expressions into Eq. 2.24.

In the injection stage, $\hat{s}_i = 0$. At the beginning where $\hat{s}_f \ll 1$, the emittance can be obtained by expanding Eq. 2.24 around $\hat{s}_f = 0$ and keeping the terms to $O(\hat{s}_f^2)$, then,

$$\epsilon_n \approx \sqrt{\sigma_{x0}^2\sigma_{px0}^2 + \frac{\bar{E}_z^2 - 8g\gamma_0(4 - 3\gamma_0)}{48\gamma_0^2}\sigma_{x0}^2\sigma_{px0}^2\hat{s}_f^2 + \frac{g^2}{12}\sigma_{x0}^4\hat{s}_f^2 + \frac{4 - 3\gamma_0^2}{12\gamma_0^2}\sigma_{px0}^4\hat{s}_f^2}$$

$$(2.25)$$

One can see the emittance grows quickly from the thermal emittance $\epsilon_{th} = \sigma_{x0}\sigma_{px0}$ at the beginning of the injection. If $\gamma_0 \approx 1$ and $\bar{E}_z \sim 1$, Eq. 2.25 can be rewritten as

$$\epsilon_n \approx \sqrt{\sigma_{x0}^2\sigma_{px0}^2 + \frac{\left(g\sigma_{x0}^2 - \sigma_{px0}^2\right)^2}{12}\hat{s}_f^2}$$

$$(2.26)$$

At the beginning, the forward velocity of the electron is not relativistic, therefore the approximations $d/dt \approx d/ds$, $\gamma \approx \gamma_0 + E_z\hat{s}$ and $F_\perp = -gr$ are not accurate at this time. Thus, Eq. 2.25 is only an indication of the initial evolution of the emittance.

As the injection continues, when $\gamma_f^{1/2} \gg 1$, Eq. 2.24 can be approximated as

$$\epsilon_n^2 \approx \frac{\left(g\gamma_0\sigma_{x0}^2 + \sigma_{px0}^2\right)^2}{3g\gamma_0}\left(1 - \sqrt{\frac{\gamma_0}{\gamma_f}} \right) - \frac{\bar{E}_z}{6g^{3/2}\gamma_0\sqrt{\gamma_f}}\left(g^2\gamma_0^2\sigma_{x0}^4 - \sigma_{px0}^4 \right)\sin\left[\frac{4\sqrt{g}\left(\sqrt{\gamma_f} - \sqrt{\gamma_0} \right)}{\bar{E}_z} \right]$$

$$(2.27)$$

From Eq. 2.27, we can see if the injection continues in many plasma skin depth ($\geq 2\pi k_p^{-1}$), the emittance would saturate at a value with small amplitude oscillation. In the transverse phase space, the first group of electrons that are ionized will have

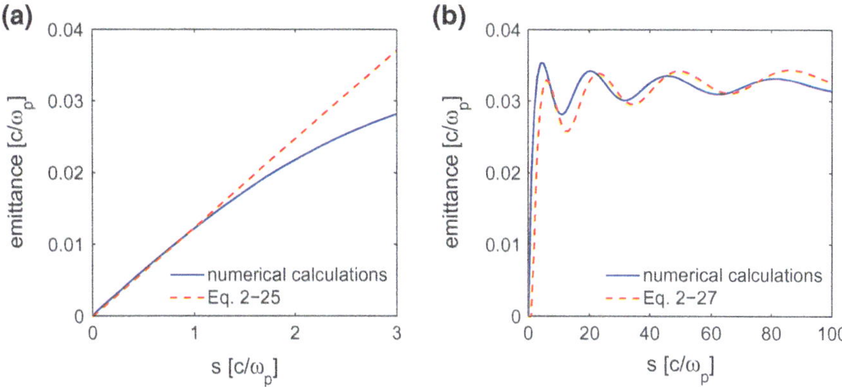

Fig. 2.9 The comparisons of the numerical calculations of Eq. 2.16 and the analytical results Eqs. 2.25 and 2.27. The injection is uniform. For simplicity, the initial distribution of the transverse phase space is $f(x, p_x) = \exp\left(-x^2/2/\sigma_x^2\right)/(\sqrt{2\pi}\sigma_x)\delta(p_x)$ with $\sigma_x = 0.3$ and $f_E(E_z) = \text{rect}\left(\frac{E_z - \bar{E}_z}{\Delta_E}\right)$ with $\Delta_E = 0.2$, $\bar{E}_z = 1.0$, and $g = 1/2$

rotated through more than an angle of π in $x - p_x$ phase space. This is illustrated in Fig. 2.8d where the trajectory in $x - p_x$ space of an electron born at rest is shown. As it is accelerated its betatron frequency and amplitude in x decreases while its amplitude in p_x increases. As a result the edge of the phase space for a collection of electrons is made up of layers of groups of electrons corresponding to each $\pi\omega_p^{-1}$ of injection. Each group corresponds to an ellipse with a different aspect ratio, and each ellipse has edge in phase space given by the trajectory shown in Fig. 2.8d. The area in phase space will therefore oscillate at twice the betatron frequency, which corresponds to the trigonometric functions in Eq. 2.27. The saturation value can be obtained from Eq. 2.27 as,

$$\epsilon_{n,sat,inj} \approx \frac{g\gamma_0\sigma_{x0}^2 + \sigma_{p_{x0}}^2}{\sqrt{3g\gamma_0}} \tag{2.28}$$

Equations 2.25 and 2.27 are examined by single particle calculations as shown in Fig. 2.9.

2.5.2 *Emittance Evolution: Decrease and Regrowth in the Acceleration Stage*

When the injection process ceases, the injected electrons rotate in the transverse phase space. The electron which is freed earlier/later with higher/lower energy rotates more slowly/fast, therefore the area occupied by the electrons in the transverse phase space decreases. With the acceleration, the difference of the acceleration gradients begins to play an important role in the betatron motion and the transverse phase space area begins to grow again until the phase ellipse determined by the focusing field in the ion column is filled in, then the emittance saturates.

In the acceleration stage, $E_z \hat{s} \gg \gamma_0$, then,

$$J(E_z, \hat{s}) \approx \begin{pmatrix} (\gamma_0/E_z\hat{s})^{1/4}\cos\phi & (g^2\gamma_0 E_z\hat{s})^{-1/4}\sin\phi \\ (g^2\gamma_0 E_z\hat{s})^{1/4}\sin\phi & (E_z\hat{s}/\gamma_0)^{1/4}\cos\phi \end{pmatrix} \tag{2.29}$$

where $\phi \approx 2\sqrt{g}\left(\sqrt{E_z\hat{s}} - \sqrt{\gamma_0}\right)/E_z$.

When $\bar{\hat{s}} \equiv (\hat{s}_f + \hat{s}_i)/2 \gg \Delta$ where Δ is the injection distance, the elements of Eq. 2.22 can be obtained as,

$$\int dE_z f_E(E_z) \int_{\hat{s}_i}^{\hat{s}_f} d\hat{s} f_s(\hat{s}) \left(J_{11}^2 \sigma_{x0}^2 + J_{12}^2 \sigma_{px0}^2\right) \approx \left[\sigma_{x0}^2 \int dE_z f_E(E_z) J_{11}^2 + \sigma_{px0}^2 \int dE_z f_E(E_z) J_{12}^2 \right.$$
$$\left. + \frac{\sigma_{x0}^2}{24} \frac{\partial^2(\int dE_z f_E(E_z) J_{11}^2)}{\partial \hat{s}^2} \Delta^2 + \frac{\sigma_{px0}^2}{24} \frac{\partial^2(\int dE_z f_E(E_z) J_{12}^2)}{\partial \hat{s}^2} \Delta^2 \right]\Bigg|_{\hat{s}=\bar{\hat{s}}}$$

$$\int dE_z f_E(E_z) \int_{\hat{s}_i}^{\hat{s}_f} d\hat{s} f_s(\hat{s}) \left(J_{11} J_{21} \sigma_{x0}^2 + J_{12} J_{22} \sigma_{px0}^2\right) \approx \left[\sigma_{x0}^2 \int dE_z f_E(E_z) J_{11} J_{21} \right.$$
$$+ \sigma_{px0}^2 \int dE_z f_E(E_z) J_{12} J_{22} + \frac{\sigma_{x0}^2}{24} \frac{\partial^2(\int dE_z f_E(E_z) J_{11} J_{21})}{\partial \hat{s}^2} \Delta^2$$
$$\left. + \frac{\sigma_{px0}^2}{24} \frac{\partial^2(\int dE_z f_E(E_z) J_{12} J_{22})}{\partial \hat{s}^2} \Delta^2 \right]\Bigg|_{\hat{s}=\bar{\hat{s}}}$$

$$\int dE_z f_E(E_z) \int_{\hat{s}_i}^{\hat{s}_f} d\hat{s} f_s(\hat{s}) \left(J_{21}^2 \sigma_{x0}^2 + J_{22}^2 \sigma_{px0}^2\right) \approx \left[\sigma_{x0}^2 \int dE_z f_E(E_z) J_{21}^2 + \sigma_{px0}^2 \int dE_z f_E(E_z) J_{22}^2 \right.$$
$$\left. + \frac{\sigma_{x0}^2}{24} \frac{\partial^2(\int dE_z f_E(E_z) J_{21}^2)}{\partial \hat{s}^2} \Delta^2 + \frac{\sigma_{px0}^2}{24} \frac{\partial^2(\int dE_z f_E(E_z) J_{22}^2)}{\partial \hat{s}^2} \Delta^2 \right]\Bigg|_{\hat{s}=\bar{\hat{s}}}$$

where the Taylor expansion is used.

Now the emittance can be expressed as

$$\epsilon_n^2 = \sigma_x^2 \sigma_{px}^2 - \sigma_{xpx}^2$$
$$\approx \sigma_{x0}^4 S_1 + \sigma_{x0}^2 \sigma_{px0}^2 S_2 + \sigma_{px0}^4 S_3 + \left(\frac{\sigma_{x0}^4}{24} S_4 + \frac{\sigma_{x0}^2 \sigma_{px0}^2}{24} S_5 + \frac{\sigma_{px0}^4}{24} S_6\right)\Delta^2 \tag{2.30}$$

where,

$$S_1 = \left(\int dE_z f_E(E_z) J_{11}^2\right)\left(\int dE_z f_E(E_z) J_{21}^2\right) - \left(\int dE_z f_E(E_z) J_{11} J_{21}\right)^2$$

$$S_2 = \left(\int dE_z f_E(E_z) J_{11}^2\right)\left(\int dE_z f_E(E_z) J_{22}^2\right) + \left(\int dE_z f_E(E_z) J_{12}^2\right)\left(\int dE_z f_E(E_z) J_{21}^2\right)$$
$$- 2\left(\int dE_z f_E(E_z) J_{11} J_{21}\right)\left(\int dE_z f_E(E_z) J_{12} J_{22}\right)$$

$$S_3 = \left(\int dE_z f_E(E_z) J_{12}^2\right)\left(\int dE_z f_E(E_z) J_{22}^2\right) - \left(\int dE_z f_E(E_z) J_{12} J_{22}\right)^2$$

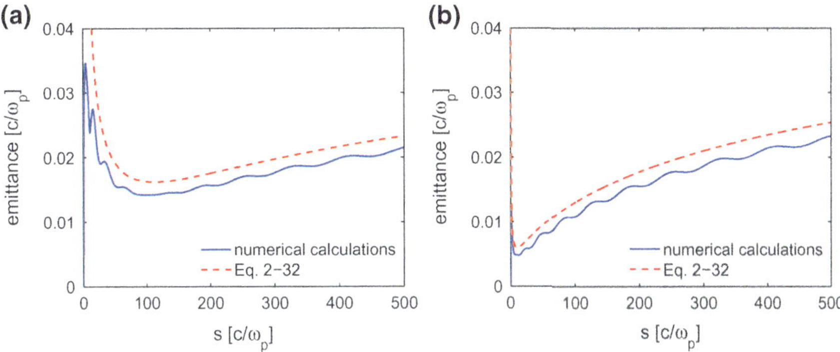

Fig. 2.10 The comparisons of the numerical calculations of Eq. 2.16 and the analytical results Eqs. 2.32. The injection is uniform in a distance **a** $\Delta = 10$ and **b** $\Delta = 1$. For simplicity, the initial distribution of the transverse phase space is $f(x, p_x) = \exp\left(-x^2/2/\sigma_x^2\right)/(\sqrt{2\pi}\sigma_x)\delta(p_x)$ with $\sigma_x = 0.3$ and $f_E(E_z) = \mathrm{rect}\left(\frac{E_z - \bar{E}_z}{\Delta_{E_z}}\right)$ with $\Delta_{E_z} = 0.2$, $\bar{E}_z = 1.0$, and $g = 1/2$

$$
\begin{aligned}
S_4 &= \left(\int dE_z f_E(E_z) J_{11}^2\right)\frac{\partial^2(\int dE_z f_E(E_z) J_{21}^2)}{\partial \hat{s}^2} + \left(\int dE_z f_E(E_z) J_{21}^2\right)\frac{\partial^2(\int dE_z f_E(E_z) J_{11}^2)}{\partial \hat{s}^2} \\
&\quad - 2\left(\int dE_z f_E(E_z) J_{11} J_{21}\right)\frac{\partial^2(\int dE_z f_E(E_z) J_{11} J_{21})}{\partial \hat{s}^2} \\
S_5 &= \left(\int dE_z f_E(E_z) J_{11}^2\right)\frac{\partial^2(\int dE_z f_E(E_z) J_{22}^2)}{\partial \hat{s}^2} + \left(\int dE_z f_E(E_z) J_{12}^2\right)\frac{\partial^2(\int dE_z f_E(E_z) J_{21}^2)}{\partial \hat{s}^2} \\
&\quad + \left(\int dE_z f_E(E_z) J_{21}^2\right)\frac{\partial^2(\int dE_z f_E(E_z) J_{12}^2)}{\partial \hat{s}^2} + \left(\int dE_z f_E(E_z) J_{22}^2\right)\frac{\partial^2(\int dE_z f_E(E_z) J_{11}^2)}{\partial \hat{s}^2} \\
&\quad - 2\left(\int dE_z f_E(E_z) J_{11} J_{21}\right)\frac{\partial^2(\int dE_z f_E(E_z) J_{12} J_{22})}{\partial \hat{s}^2} \\
&\quad - 2\left(\int dE_z f_E(E_z) J_{12} J_{22}\right)\frac{\partial^2(\int dE_z f_E(E_z) J_{11} J_{21})}{\partial \hat{s}^2} \\
S_6 &= \left(\int dE_z f_E(E_z) J_{12}^2\right)\frac{\partial^2(\int dE_z f_E(E_z) J_{22}^2)}{\partial \hat{s}^2} + \left(\int dE_z f_E(E_z) J_{22}^2\right)\frac{\partial^2(\int dE_z f_E(E_z) J_{12}^2)}{\partial \hat{s}^2} \\
&\quad - 2\left(\int dE_z f_E(E_z) J_{12} J_{22}\right)\frac{\partial^2(\int dE_z f_E(E_z) J_{12} J_{22})}{\partial \hat{s}^2}
\end{aligned}
\tag{2.31}
$$

Assuming $f_E(E_z) = \mathrm{rect}\left(\frac{E_z - \bar{E}_z}{\Delta_E}\right)$ and $\bar{E}_z \gg \Delta_E$, we can get the integrations over E_z and only keep the dominant terms. After a lengthy derivations as shown in Appendix A, the emittance is obtained as

$$
\epsilon_n \approx \frac{1}{2\sqrt{g\gamma_0}}\sqrt{\left(g\gamma_0\sigma_{x0}^2 + \sigma_{px0}^2\right)^2 - \left(g\gamma_0\sigma_{x0}^2 - \sigma_{px0}^2\right)^2\left[\frac{\bar{E}_z}{g\hat{s}} - \frac{1}{3}\left(\frac{\Delta}{\hat{s}}\right)^2\right]\left(\frac{\bar{E}_z}{\Delta_E}\right)^2 \sin^2\left(\sqrt{\frac{g\bar{s}}{\bar{E}_z}}\frac{\Delta_E}{\bar{E}_z}\right)}
\tag{2.32}
$$

We check the analytical formula Eq. 2.32 with single particle calculations as shown in Fig. 2.10 and good agreements are obtained.

From Eq. 2.32, one can find there is a position where the emittance achieves its minimum value. This position can be obtained by solving a transcendental equation as

$$\tan\left(\sqrt{\frac{g\bar{\hat{s}}}{\bar{E}_z}}\frac{\Delta_E}{\bar{E}_z}\right)\left(\sqrt{\frac{g\bar{\hat{s}}}{\bar{E}_z}}\frac{\Delta_E}{\bar{E}_z}\right)^{-1} = 1 + \frac{g\Delta^2}{3\bar{E}_z\bar{\hat{s}} - 2g\Delta^2} \qquad (2.33)$$

Over long distance acceleration, the emittance finally saturates around

$$\epsilon_{n,sat} = \frac{g\gamma_0\sigma_{x0}^2 + \sigma_{px0}^2}{2\sqrt{g\gamma_0}} \geq \epsilon_{th} = \sigma_{x0}\sigma_{px0} \qquad (2.34)$$

One can see the saturation emittance is determined by the sum of squares of the initial spot size σ_{x0} and the residual momentum σ_{px0}, which is typically larger than the thermal emittance $\epsilon_{th} = \sigma_{x0}\sigma_{px0}$. If initially $\sqrt{g\gamma_0}\sigma_{x0} = \sigma_{px0}$, the saturation emittance will be the same as the thermal emittance, and this indicates a matching condition for constant emittance. In the limiting case with $\Delta = 0$ and $\Delta_E = 0$, Eq. 2.32 is simplified as $\epsilon_n = \epsilon_{th}$, i.e., the emittance stays as the thermal emittance if all electrons are injected at the same time and feel the same acceleration gradient.

As shown above, the transverse phase mixing increases the ultimate emittance to a value that could be much larger than initial thermal emittance. Therefore one way to reduce the emittance is to suppress the transverse phase mixing by limiting the injection distance. When the injection distance is on the order of or shorter than the plasma skin depth, i.e., $\Delta \lesssim 1$, the emittance growth in the injection stage is strongly suppressed, as shown in Fig. 2.10b where the initial emittance with $\Delta = 1$ is several times lower than the initial emittance with $\Delta = 10$ in Fig. 2.10a. This idea has been ideally realized in the transverse injection scheme in where two transversely colliding laser pulses are used to release the electrons. The intensity of a single laser is not high enough to release the electrons, after they overlap, the combined field is high enough to ionize the electrons. Thus, the injection distance is on the order of the spot size of the lasers which can be much smaller than the plasma skin depth. The ultimate emittance in this scheme can be as low as 10 nm [23].

2.5.3   A Phenomenological Model

Besides the above lengthy derivation, a simple phenomenological model can also be used to describe the emittance evolution in certain specific cases. If the residual momentum is negligible, i.e., the electrons are born at rest, as shown in Fig. 2.8 the emittance can be determined by two quantities, one is the saturation emittance and the other is the phase spread occupied by the beam, i.e., $\phi_M - \phi_m$, where ϕ_M (ϕ_m) represents the maximum (minimum) betatron phase occupied by the particles.

When the electrons are born at rest, i.e., $p_{x0} = 0$, then for a given time, from Eq. 2.20 we can get

$$x = x_0 \left(\frac{\gamma_0}{\bar{\gamma}}\right)^{1/4} \cos\phi$$

$$p_x = x_0 \left(g^2\gamma_0\bar{\gamma}\right)^{1/4} \sin\phi \tag{2.35}$$

where the $\bar{\gamma}$ is the average energy of the injected electrons at this time, $\phi = 2\sqrt{g}(\sqrt{\bar{\gamma}} - \sqrt{\gamma_0})/E_z$ is the betatron phase. Because the amplitudes of the phase space coordinates have a weak dependence on the particle energy ($\propto \gamma^{1/4}$), we use the average energy $\bar{\gamma}$ to replace the energy for each particle in the oscillation amplitudes. Then, we can obtain

$$\sigma_x^2 = \frac{\sigma_{x0}^2}{2}\left(\frac{\gamma_0}{\bar{\gamma}}\right)^{1/2}\left[1 + \frac{\sin2\phi_M - \sin2\phi_m}{2\phi_M - 2\phi_m}\right]$$

$$\sigma_{p_x}^2 = \frac{g\sigma_{x0}^2}{2}\sigma_{x0}^2\left(\gamma_0\bar{\gamma}\right)^{1/2}\left[1 - \frac{\sin2\phi_M - \sin2\phi_m}{2\phi_M - 2\phi_m}\right]$$

$$\sigma_{xp_x} = \frac{\sqrt{g}}{2}\sigma_{x0}^2\gamma_0^{1/2}\frac{\cos2\phi_M - \cos2\phi_m}{2\phi_M - 2\phi_m} \tag{2.36}$$

where $\sigma_{x0}^2 = \int dx_0 x_0^2 f_0(x_0)$ and $f_0(x_0)$ is the normalized distribution function when the electrons are born. Then, the normalized emittance

$$\epsilon_n(\Delta\phi) = \sqrt{\sigma_x^2\sigma_{p_x}^2 - \sigma_{xp_x}^2}$$

$$= \epsilon_{n,sat}\sqrt{1 - \left(\frac{\sin\Delta\phi}{\Delta\phi}\right)^2} \tag{2.37}$$

where $\Delta\phi = \phi_M - \phi_m$. Note that the emittance expression derived from this phenomenological model Eq. 2.37 has a similar form of the expression derived from the general method Eq. 2.32. And the saturation emittance when the phase ellipse is filled out is

$$\epsilon_{n,sat} = \frac{\sqrt{g\gamma_0}}{2}\sigma_{x0}^2 \tag{2.38}$$

One can see the saturation emittance derived in this phenomenological model is the same as the expression Eq. 2.34 if $\sigma_{p_{x0}} = 0$.

During the injection process, i.e., $\phi_m = 0$, we can neglect the spread of the acceleration gradient, then,

$$\Delta\phi = \phi_M \approx 2\sqrt{g}\frac{\sqrt{\gamma_0 + \bar{E}_z\hat{s}} - \sqrt{\gamma_0}}{\bar{E}_z} \tag{2.39}$$

At the beginning of the injection, i.e., $\Delta\phi < 1$, by using Taylor expansion, we can get

$$\epsilon_n \approx \epsilon_{n,sat} \frac{\Delta\phi}{\sqrt{3}} \approx 2\epsilon_{n,sat} \sqrt{\frac{g}{3}} \frac{\sqrt{\gamma_0 + \bar{E}_z \hat{s}} - \sqrt{\gamma_0}}{\bar{E}_z} \tag{2.40}$$

where the average acceleration gradient $\bar{E}_z$ is used in the above expressions.

After the injection ceases, for each injected electron $\phi \gg 1$ and $\delta\phi \equiv \phi - \langle\phi\rangle \ll \phi$, leading to

$$\frac{\delta\phi}{\phi} \approx \frac{1}{2}\left(\frac{\delta\hat{s}}{\hat{s}} - \frac{\delta E_z}{E_z}\right) \tag{2.41}$$

The variance of $\delta\phi$ can be obtained by assuming the independence between the accelerating gradient and the injection time,

$$\sigma_\phi \equiv \sqrt{\langle\delta\phi^2\rangle} \approx \sqrt{\frac{g}{E_z}\left[\frac{\sigma_s^2}{\hat{\bar{s}}} + \bar{\hat{s}}\left(\frac{\sigma_{E_z}}{\bar{E}_z}\right)^2\right]} \tag{2.42}$$

where $\sigma_s^2 = \langle(\hat{s}_i - \langle\hat{s}_i\rangle)^2\rangle$ and $\sigma_{E_z}^2 = \langle(E_z - \langle E_z\rangle)^2\rangle$. To obtain an expression of $\Delta\phi$ in terms of σ_ϕ, a certain distribution of electrons need to be assumed. For a uniform distribution, $\Delta\phi = \sqrt{12}\sigma_\phi$, Eq. 2.37 can be rewritten as

$$\epsilon_n = \epsilon_{n,sat}\sqrt{1 - \left(\frac{\sin\sqrt{12}\sigma_\phi}{\sqrt{12}\sigma_\phi}\right)^2} \tag{2.43}$$

Equations 2.42 and 2.43 predict that for $\bar{\hat{s}} < (\bar{E}_z/\sigma_{E_z})\sigma_s \equiv \hat{s}_0$ the emittance actually decreases; it reaches a local minimum at $\hat{s}_0$, and then increase until it saturates at $\epsilon_{n,sat}$.

2.5.4 Comparisons with PIC Simulations

We next compare these predictions against self-consistent 2D OSIRIS simulations of injection triggered by a laser pulse into the wake produced by an electron beam. As schematically shown in Fig. 2.11a, in the simulation the beam driver propagates in a mixture of a plasma with a density $n_p = 1.6 \times 10^{17}$ cm^{-3} and neutral He gas of density $n_{He} = 1.6 \times 10^{13}$ cm^{-3}. Due to small electric field of the beam driver, the helium atoms remain neutral after the beam passes. Therefore only a low intensity laser is required to ionize the helium, leading to a small residual momentum. In the sample simulations the laser is co-propagating with the wake. The laser is tightly

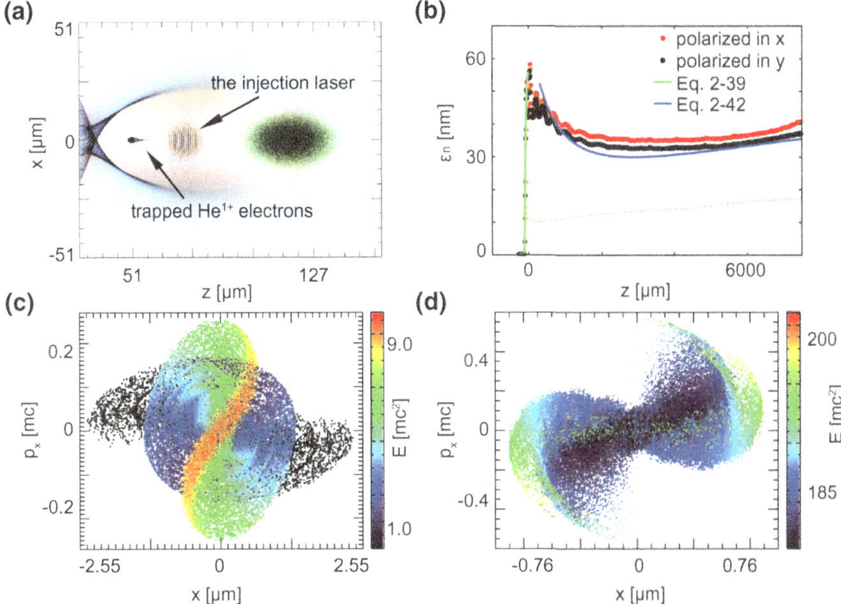

Fig. 2.11 **a** 2D PIC simulation of ionization injection by a laser into a beam driven wake. Drive beam (green): $\sigma_r = 6.4$ μm, $\sigma_z = 10.6$ μm, $n_b = 2.6 \times 10^{17}$ cm^{-3}, $E_b = 2$ GeV. Laser: $\lambda = 800$ nm, $a_0 = 0.04$, $w_0 = 3$ μm, $\tau \approx 30$ fs. **b** Comparison of the emittance evolution between simulations and theory. The red and black lines are for the laser polarized in or out of the simulation plane respectively. The green dotted line shows the emittance evolution for limited helium range (19 μm $\approx 1.5 k_p^{-1}$). The $x - p_x$ phase space when the injection is terminated (**c**) and when the emittance achieves its minimum (Reprinted from Ref. [55]. Copyright with kind permission from American Physical Society)

focused to reduce the initial spot size when the electrons are ionized. The reduced initial spot size and reduced residual momentum lead to a much reduced emittance.

We adopt the convention that the beam drive propagated in the $\hat{z}$ direction where the other direction in the simulation plane is the $\hat{x}$ direction. In some cases, the injection laser is polarized in the simulation plane (along $\hat{x}$) and in other cases not in the simulation plane (along $\hat{y}$). We label the emittance in the $x - p_x$ plane from a simulation with a subscript x and y depending on the polarizing direction. The simulation window has a dimension of 508×635 μm with 4000×5000 cells in the $\hat{x}$ and $\hat{z}$ directions, respectively. This corresponds to cell sizes of $0.2 k_0^{-1}$ in the $\hat{x}$ and $\hat{z}$ directions, where k_0 is the wavenumber of the injection laser. The simulation uses 2 and 4 particles per cell for the plasma and neutral He respectively.

In Fig. 2.11b the evolution of the emittance in the simulation and the predictions of Eq. 2.39 (during ionization) and Eq. 2.42 (after ionization) using σ_z and σ_{E_z} calculated from the simulation are compared, and the agreement is very good. In this co-propagating configuration, the ionization duration is limited by the Rayleigh length of the laser and is ~ 0.6 ps $> \pi \omega_p^{-1}$. Therefore there are several oscillations in the injection stage as shown in Fig. 2.11b. In Fig. 2.11c we show $x - p_x$ at the end

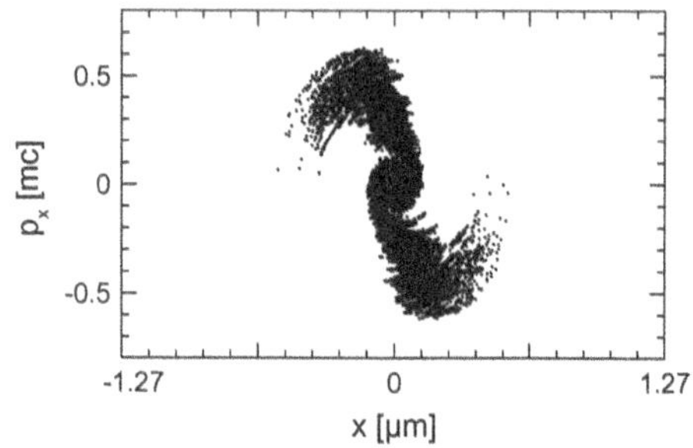

Fig. 2.12 The transverse phase space for limited helium range (19 μm $\approx 1.5k_p^{-1}$) at $z = 5.2$ mm. We can see the transverse phase mixing is strongly suppressed

of the injection. One can see the first group electrons have rotated near a period in the phase space. In Fig. 2.11d we show $x - p_x$ for $\hat{s} = \hat{s}_0$ where the emittance reaches its minimum, and it can be clearly seen that the range of $\Delta\phi < \pi/2$ at this time even though $\Delta\phi \approx 3\pi/2$ at the end of injection. We also show the emittance evolution for limited helium range (19 μm $\approx 1.5k_p^{-1}$) by the green dotted line in Fig. 2.11b. One can see the emittance is lower than 20 nm within 7.5 mm acceleration distance. The transverse phase space at $z = 7.5$ mm is shown in Fig. 2.12. One can see the beam remains as a narrow leaf which indicates the transverse phase mixing is strongly suppressed.

2.6 Longitudinal Phase Mixing

The above transverse phase mixing process is quite similar to certain phenomena in traditional accelerator physics. For example, in a photoinjector, the transverse motion of the electron beam is affected by the self space charge field, therefore different slice with different current feels different space charge force, leading to a different betatron phase. As a result the projected emittance increases while the slice emittance stays constant. This kind of emittance growth can be reverted by carefully designed external focusing field [56]. In ionization injection, the physics of transverse phase mixing is very different in the sense that it occurs in every slice due to a longitudinal phase mixing process. When the electrons are released, they slip back to the tail of the wake. The final longitudinal position of the electron in the wake is determined by its initial position and the wake structure, as implicitly expressed in the trapped condition $\delta\psi \approx -1$. Due to the quasi-static nature of the injection process, the electrons ionized at different times can reside in the same longitudinal slice, therefore this longitudinal phase mixing leads to a transverse phase mixing in each slice.

2.6.1 The Trapping Condition

In the quasi-static approximation, i.e., the dependence on the longitudinal position z and the time t can be combined as $\xi = z - v_\phi t$, it is straightforward to show that

$$\gamma - v_\phi p_z - \psi = \text{const} \tag{2.44}$$

When an electron gets trapped, i.e., $v_z = v_\phi$, we can get

$$p_z = \frac{1 + p_\perp^2 - (\lambda + \delta\psi)^2}{v_\phi(\lambda + \delta\psi)} \tag{2.45}$$

$$\gamma = \frac{1 + p_\perp^2}{\lambda + \delta\psi} \tag{2.46}$$

where $\lambda = \gamma_0 - v_\phi p_{z0}$. For an electron starts at rest, $\lambda = 1$.

After a straightforward algebra, we can get

$$\delta\psi = -\lambda + \sqrt{\frac{1 - v_\phi^2}{1 - v_\phi^2 - v_\perp^2} \frac{1}{\gamma_\phi}} \tag{2.47}$$

where the negative branch is neglected because $\lambda + \delta\psi = \gamma - v_\phi p_z > 0$.

The phase velocity of the wake is equal to the group velocity of the driver. For a laser driver, $\gamma_\phi \approx \omega_0/\omega_p$, which is on the order of a few tens; for a beam driver, $\gamma_\phi = \gamma_b$, which is typically much larger than the laser driver case. Thus, the trapping condition Eq. 2.47 can be approximated as $\delta\psi \approx -1$.

2.6.2 Longitudinal Phase Mixing

The time scale for the driver evolution is typically much slower than the injection process, therefore the quasi-static approximation can be used for injection. By applying the trapping condition we get

$$\psi(\xi_f, r_f) - \psi(\xi_i, r_i) \approx -1 \tag{2.48}$$

The final relative longitudinal position only depends on the initial relative longitudinal position and the transverse positions. From the definition of the relative longitudinal position $\xi \equiv z - v_\phi t$, the longitudinal slice of the injected beam can contain electrons originated from different times and different initial longitudinal positions.

For a sufficient large bubble (the maximum radius $r_m > 2$), the nonlinear plasma wakefield theory shows the pesudo-potential in the ion column can be expressed as [53, 54].

$$\psi(\xi, r) \approx \frac{r_b^2(\xi) - r^2}{4} \tag{2.49}$$

and

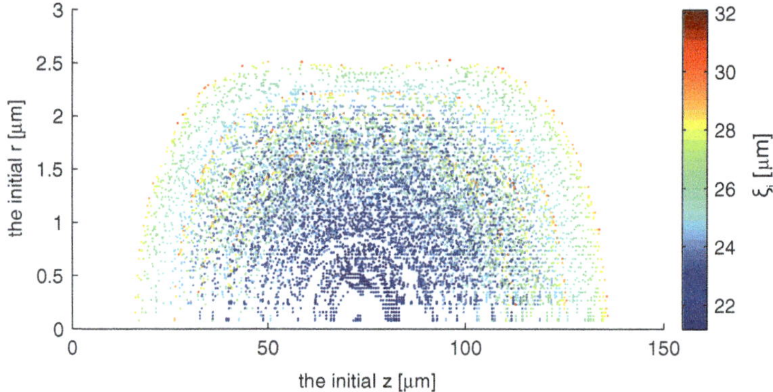

Fig. 2.13 3D OSIRIS simulation result: the dependence of ξ_i on r_i and z_i. A 800 nm laser and neutral helium are initialized in the simulation. The laser parameters: $a_0 = 0.04$, $w_0 = 4\,\mu$m, $\tau = 18$ fs. The focal plane is $z = 76\,\mu$m. Note that the ξ_i here only has relative meaning

$$r_b^2(\xi) \approx r_m^2 - \xi^2 \tag{2.50}$$

where $r_b(\xi)$ is the radius of the ion channel.

Then,

$$\xi_f \approx \sqrt{4 + \xi_i^2 + r_i^2 - r_f^2} \tag{2.51}$$

From Eq. 2.21, $r \propto \gamma^{-1/4}$, therefore r_f is much smaller than r_i and if $r_i \ll 1$, the final relative longitudinal position can be simplified as

$$\xi_f \approx \sqrt{4 + \xi_i^2} \tag{2.52}$$

For a gaussian laser beam, the field envelop is

$$E_x = E_0 \left[\frac{w_0}{w(z)}\right]^2 \exp\left[-\frac{r^2}{2w(z)^2}\right] \exp\left[-\frac{(t-z)^2}{2\tau^2}\right] \tag{2.53}$$

where $w(z) = w_0\sqrt{1 + (z/z_R)^2}$ is the spot size, $z_R = \pi w_0^2/\lambda_0$ is the Rayleigh length of the laser. We can obtain the dependence of the initial relative longitudinal position ξ_i on z_i, t_i and the transverse position r_i from Eq. 2.53. The electrons near the focal plane is ionized with smaller ξ_i compared with the electron far away from the focal plane, the electrons with larger r_i is ionized with smaller ξ_i compared with the electrons with smaller initial radius. This is confirmed by 3D PIC simulations as shown in Fig. 2.13.

We also analyze the simulation data in Sect. 2.5.4 to check Eq. 2.52. In Fig. 2.14a we plot the relation between ξ_f and ξ_i at a propagation distance of $z \approx 1$ mm and

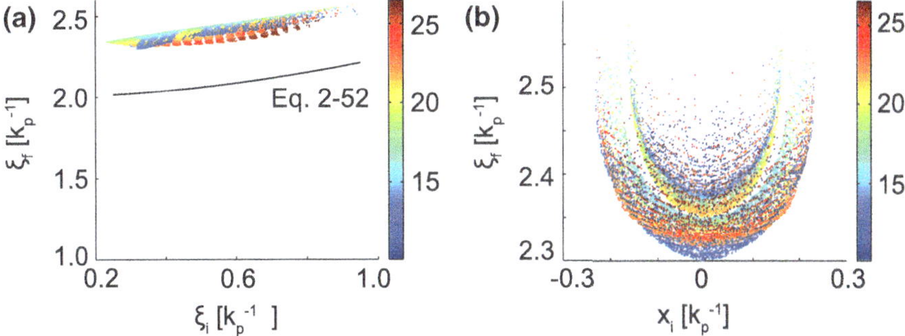

Fig. 2.14 **a** The dependence of ξ_f on ξ_i from 2D simulation and comparison with Eq. 2.52. **b** The dependence of ξ_f on x_i. Note that different colors represent different electron birth times (Reprinted from Ref. [55]. Copyright with kind permission from American Physical Society)

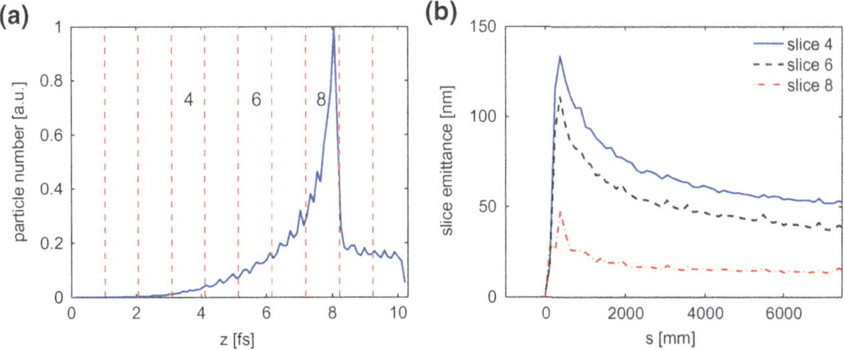

Fig. 2.15 **a** The injected electrons are divided into 10 slices. **b** The slice emittance evolution in slice 4, 6 and 8

compare with Eq. 2.52, and a similar trend is seen. In Fig. 2.14b we plot the relation between ξ_f and initial $r = x$ of each electron, with each color representing a different birth time. From the color code of both Fig. 2.14a, b, electrons born at the same time are seen to be distributed into each slice.

To check the transverse phase mixing in each slice, we divide the injected electrons into 10 slices and plot the slice emittance evolution in Fig. 2.15. The slice emittance evolution has similar behavior with the projected emittance evolution, i.e., rapid growth at the beginning of the ionization and decrease in the acceleration stage. However the spread of the acceleration gradient in a slice is much smaller than the spread of the whole beam, therefore it takes a much longer distance for the emittance to regrow in the acceleration stage which can not be seen in the limited simulation time.

Due to the longitudinal phase mixing, a longitudinal slice contains particles ionized at different times, therefore the slice energy spread is determined by the injection distance, i.e., $\Delta\gamma \approx E_z\Delta_{slice}$, where Δ_{slice} is the injection distance of each slice. Reducing the injection distance can lead to a much reduced slice energy spread.

2.7 Space Charge Effects

The single particle model described in previous sections does not take into account the self space charge forces between the injected particles. In plasma based accelerators, the acceleration gradient is very high, typically on the order of GV/cm, therefore the electrons can be accelerated to relativistic energy within tens of μm. However the density of the injected electrons could also be very high, quite often much larger than the background plasma density. This indicates the self space charge forces could play an important role in the evolution of the beam phase space.

For relatively large injected charge, the nonlinear space charge force of the injected electrons in both the longitudinal and transverse directions could modify the trajectories of the electrons while they are at low energy, thereby modifying the evolution of the emittance. For cases where the total injection phase is larger than π, space charge effects lead directly to a saturation of the emittance around ϵ_{sat}. In Fig. 2.16a this saturation behavior can be clearly seen. The quadratic dependence of ϵ_{sat} on the laser spot size is also verified through PIC simulations with a good agreement (Fig. 2.16b). The evolution of the slice emittance in high charge case is shown in Fig. 2.17. One can see the slice emittance evolution has similar behavior as the projected emittance.

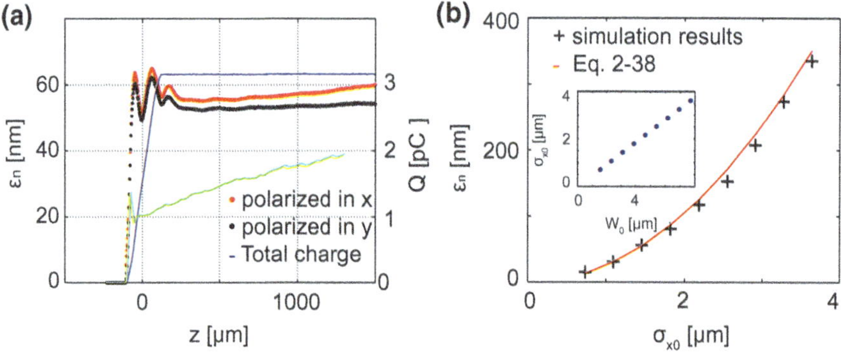

Fig. 2.16 **a** Emittance evolution with space charge effect. The red and black lines show the emittance evolution for the laser in or out of the 2D simulation plane respectively with $n_{He} = 1.6 \times 10^{16}$ cm^{-3}. The green line shows the emittance evolution for limited Helium range (19 μm), $n_{He} = 1.6 \times 10^{17}$ cm^{-3}. The charge value (the blue curve) is estimated by assuming the beam is symmetric in 3D. **b** The quadratic dependence of the saturated emittances on σ_{x0} for large injection phase ($> \pi$) with space charge effect. The laser intensity was fixed at $a_0 = 0.04$ for different spot sizes, and the inner plot shows the dependence of σ_{x0} on the laser spot size (Reprinted from Ref. [55]. Copyright with kind permission from American Physical Society)

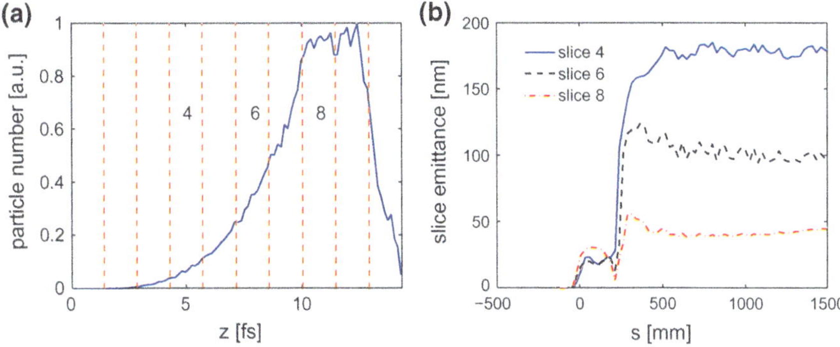

Fig. 2.17 **a** The injected electrons are divided into 10 slices. **b** The slice emittance evolution in slice 4, 6 and 8

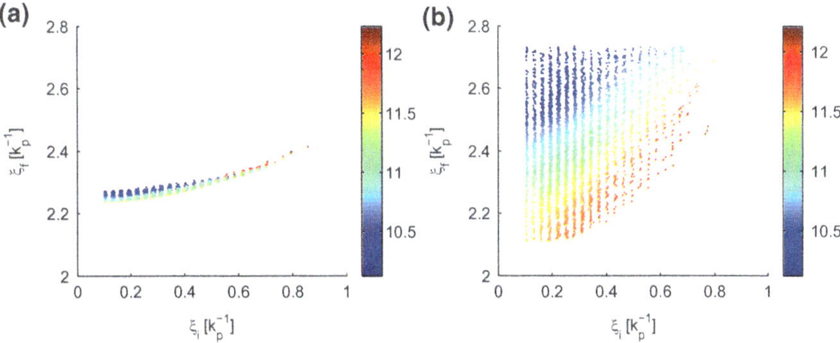

Fig. 2.18 The dependence of ξ_f on ξ_i when the helium is limited in 19 μm $\approx 1.5k_p^{-1}$. **a** $n_{He} = 10^{-4}n_p$, **b** $n_{He} = n_p$. The color represent the ionization time

In the transverse injection, due to small injection volume, the initial density of the trapped electrons is much higher than the density in the longitudinal injection in order to obtain the same charge. In the simulation shown in Ref. [23], the helium density is even higher than the background plasma density. Therefore, the self space charge force play a very important role in the phase space dynamics of the trapped electrons. At the beginning of the ionization, the electrons are non-relativistic, the self space charge force may blow out the electrons, leading to a longer pulse duration. A comparison for limited injection distance with different helium density is shown in Fig. 2.18. One can see the relation of ξ_f and ξ_i (Eq. 2.52) is broken by the self space charge force when $n_{He} = n_p$. In this case, the longitudinal phase mixing is much reduced and a very low slice energy spread electron beam (tens of keV) can be generated as shown in Fig. 2.19.

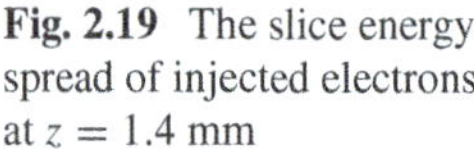

Fig. 2.19 The slice energy spread of injected electrons at $z = 1.4$ mm

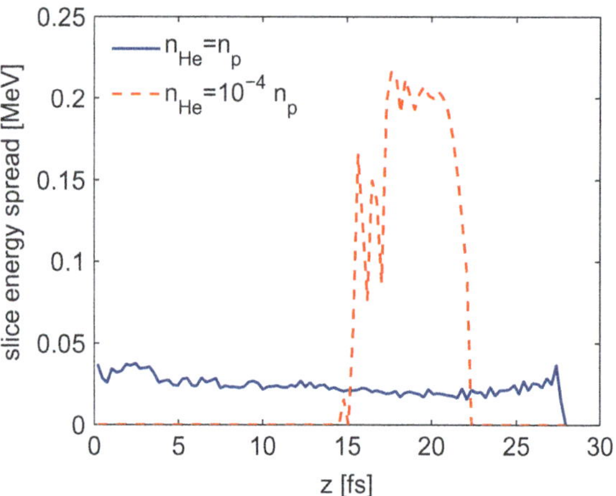

2.8   The Two-Color Ionization Injection

As shown in previous sections, the emittance of the injected beam in the single laser ionization injection scheme is on the order of μm in the two transverse directions due to the large residual momentum and the large initial spot size. An obvious way to reduce the intensity and the spot size of the laser is to sperate the wake excitation process and the injection process using multi pulses. This strategy has been demonstrated recently in two schemes utilizing an electron beam driver to excite the wake. In this section, we propose a high quality electron beam generation scheme using two or more laser pulses with different wavelengths. In this all optical concept, two laser pulses with different wavelengths—a long wavelength ($\sim$10 μm) laser pulse with a large ponderomotive potential but small intensity is used to drive the wake and a short wavelength ($\sim$400 nm) laser pulse with a small ponderomotive potential but a large intensity is used to inject electrons into the wake. Since a long wavelength laser pulse has a large ponderomotive potential, it still produces a large amplitude wake at a relatively low intensity. This is analogous to the beam-driven wake where a much lower beam electric field can be used to drive a strong wake. Our PIC simulation results show that it is possible to generate electron beams containing $\sim$4 pC of charge with small normalized emittances ($\sim$50 nm), more than 1 order of magnitude smaller than that obtained in ionization injection using a single 800 nm wavelength laser. With the transverse colliding geometry, simulations show that similarly low emittance and much lower slice energy spread ($\sim$30 keV, comparing with the typical value of few MeV in the longitudinal injection scheme) can be simultaneously obtained for electron beams with a few pC charge. Such low slice energy spread may have significant advantages in applications relevant to future coherent light sources driven by plasma accelerators.

2.8.1 The Emittance in a Single Laser Case

In this subsection we study the ultimate beam emittance achieved when the wake is both excited and the charge is injected via ionization injection by a single laser pulse. At the beginning, we present results from simulation A in Table 2.2. Figure 2.20a shows a typical scenario for ionization injection using a single laser pulse, where an intense short pulse laser propagates through a mixture of helium and nitrogen gases. The simulation is initialized with the two electrons from helium atoms being fully ionized and the first five L-shell electrons from nitrogen atoms are stripped off at the very leading edge of the laser pulse and are pushed out by the ponderomotive force of the laser, thereby forming the nonlinear wake. The two K-shell electrons of nitrogen atom have very high ionization potentials (552 and 667 eV respectively), therefore they can only be freed in the middle of the wake near the peak of the laser intensity (e.g., 10^{19} W/cm^2 or $a_0 \sim 2$ for 800 nm laser wavelength). This is evident in the plots of the ionization levels, the laser electric field, and the wake potential shown in Fig. 2.20b. After ionization, these K-shell electrons start to gain energy in the wake while still slipping backwards relative to the wake until their longitudinal velocity approaches the phase velocity of the wake. After this point, they are trapped in the wake and then keep gaining energy until they outrun the wake by dephasing.

The size of the ionization volume depends on both the intensity profile and the spot size of the laser, and it is typically a small fraction of the laser spot size. In the simulation shown in Fig. 2.20, the rms size of this column σ_{x0} is 2.7 μm, about one fifth of the laser spot size w_0. The residual transverse momenta in the $\hat{x}$ direction are $\sigma_{p_{0,x}} = 0.65$ mc and $\sigma_{p_{0,y}} = 0.41$ mc respectively, where x and y refer to cases where the laser is polarized in ($\hat{x}$) or out ($\hat{y}$) of the simulation plane. The momenta for the $\hat{x}$ case is from both the 1D residual drift and the ponderomotive force while in the $\hat{y}$ case it is only from the ponderomotive force. Therefore, for this example both effects are important, but the ponderomotive acceleration is more important.

In Fig. 2.20c, we plot the emittance evolution from simulation A in Table 2.2. The initial growth and saturation of the beam emittance in the $x - p_x$ plane can be clearly observed. The emittance eventually saturates at a value approximately given by Eq. 2.34. For ionization injection by a single laser pulse, the typical value of emittance is about 1 μm along the laser polarization direction (the laser was polarized in $\hat{x}$ direction) and few hundreds nm perpendicular to the laser polarization (the laser was polarized in the $\hat{y}$ direction). From the previous analysis, we know the saturated emittance ϵ_{sat} is determined by the values of σ_{x_0} and $\sigma_{p_{x0}}$. Therefore, developing concepts to reduce both of these values could lead to lower final emittance.

For the single laser pulse scheme of ionization injection, one obvious possibility for reducing σ_{x_0} and $\sigma_{p_{x0}}$ is to decrease the laser intensity. However the trapping condition for electron injection $\delta\psi \approx -1$ places a lower limit on the wake amplitude and hence a lower limit on the laser intensity. The electrons can be trapped if $\psi_{min} - \psi_{birth} < -1$, where ψ_{min} is the minimum value of the pseudo potential and ψ_{birth} (electrons can be born off axis) is the pseudo potential where the electrons are released. Figure 2.21 shows the pseudo potentials for different driver intensities.

Table 2.2 Simulation parameters of the presented results in Sect. 2.8

Simulation parameters	A	B	C	D	E	F
Box size (k_0^{-1})	[450, 500]	[350, 500]	[250, 500]	[300, 400]	[400, 400]	[400, 400]
Number of cells	[4500, 1600]	[1750, 1000]	[1250, 2500]	[9000, 800]	[50000, 800]	[2000, 20000]
Time step (ω_0^{-1})	0.05	0.1	0.1	0.033	4e-3	1.99e-2
Particles per cell: plasma electrons	[1, 1]	[1, 1]	[1, 1]	[1, 1]	[1, 1]	[1, 1]
Particles per cell: ions	[4, 4]	[2, 2]	[4, 4]	[4, 4]	[2, 2]	[2, 2]
Preformed plasma density $(10^{17}\,\mathrm{cm}^{-3})$	21	21	21	17	0.11	0.11
Dopant gas species	Nitrogen	Carbon	Nitrogen	Nitrogen	Oxygen	Oxygen
Dopant gas atomic density $(10^{17}\,\mathrm{cm}^{-3})$	0.21	0.21	0.21	3.4	0.022	0.022
Laser wavelengths $(\mu\mathrm{m})$: λ_0, λ_1	0.8	0.8	0.8, 0.8	0.8, 0.08	10, 0.4	10, 0.4
Laser amplitudes: a_0, a_1	2.0	1.4	1.2, 2.0	1.2, 0.25	1.4, 0.09	1.4, 0.06
Laser spot sizes $(\mu\mathrm{m})$: w_0, w_1	13	13	11, 2.5	10, 0.64	142, 4.8	142, 48
Laser pulse lengths (fs): τ_0, τ_1	32	51	32, 12	18, 7	352, 44	352, 13
Laser pulse delay (fs)	None	None	40	30	584	245
Saturation emittances (nm): $\epsilon_\parallel, \epsilon_\perp$	1200, 350	1500, 500	1600, 800	20, 8	50, 50	60

The normalized vector potential of the laser a_0 is varied while the laser intensity profile was kept the same as in the simulation A. It is found that the minimum a_0 for $(\delta\psi)_{min} \approx -1.0$ is about 1.3. We note that to use this intensity to reach injection, carbon needs to be chosen instead of nitrogen to provide the K-shell electrons

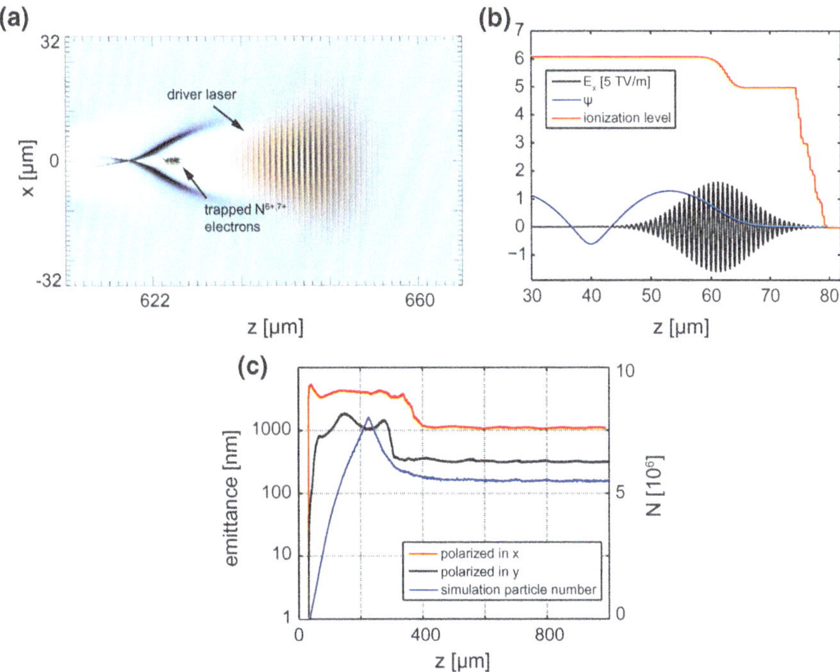

Fig. 2.20 2D OSIRIS simulation (A) of wake excitation and ionization injection driven by a single laser pulse. Laser parameters: $\lambda_0 = 800$ nm, $a_0 = 2$, $w_0 = 13\,\mu$m and $\tau_{FWHM} = 32$ fs (Gaussian profile). Plasma profile: Total background electrons $n_e = 2.1 \times 10^{18}$ cm^{-3}, include the pre-ionized plasma electrons and the five L-shell nitrogen electrons. The N^{5+} ion has $n_{N5+} = 0.01 n_e$ and a limited range (190 μm). **a** Snapshot of the charge density distribution of the wake electrons, the K-shell electrons of nitrogen, and the electric field in x direction. **b** The laser electric field E_x, the normalized pseudo potential ψ of the plasma wake and the ionization state of nitrogen atoms on axis. **c** The emittance evolution of the beam along (red) and perpendicular to the polarization direction (black), and the evolution of the particle number in the simulation (blue). The saturated emittance are 1200 nm ($\hat{x}$) and 350 nm ($\hat{y}$) for the laser polarized in and out of the simulation plane respectively. Due to the particle loss as shown in the blue line, the saturated emittance is smaller than the thermal emittance which contains all the ionized particles. In this simulation the injection distance is controlled by the longitudinal nitrogen distribution. The injected charge is about 40 pC by assuming a cylindrical distribution around z-axis (Reprinted from Ref. [57]. Copyright with kind permission from American Physical Society)

for trapping due to their relatively lower IP than that of nitrogen and the ionization needs to occur at the maximum of ψ. In Fig. 2.22, a 2D PIC simulation (simulation B) for $a_0 = 1.4$ is shown to confirm that the injection does occur at this low intensity. However the emittances obtained in this case, $\sim$1500 nm($\hat{x}$) and $\sim$500 nm($\hat{y}$), are not better than those obtained in the simulation A. This indicates that it is difficult to further reduce the emittance by just fine tuning the laser parameters of the single laser pulse for ionization injection.

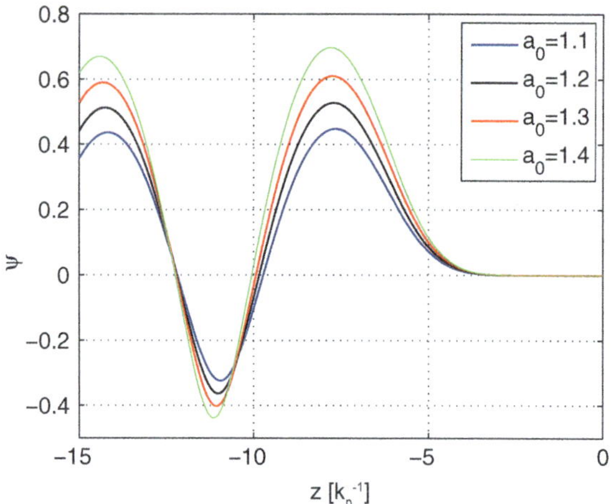

Fig. 2.21 The normalized pseudo potential ψ of the plasma wake on axis for different a_0 while keeping the intensity profile of the laser pulse fixed (Reprinted from Ref. [57]. Copyright with kind permission from American Physical Society)

In this simulation the $(\delta\psi)_{min}$ is close to -1, therefore only electrons with large initial transverse positions which lead to large transverse momenta when they cross the z-axis are injected (Eq. 2.47: $\delta\psi < -1 + \frac{1}{\gamma_\phi}\sqrt{\frac{1-v_\phi^2}{1-v_\phi^2-v_\perp^2}}$) and form a phase space ring structure in 2D simulations. The $x - p_x$ phase space plots are shown in Fig. 2.22c.

2.8.2 The Two-Color Ionization Injection: Longitudinal Injection

As mentioned earlier, in order to generate electron beams with emittances below 100 nm, the transverse size of the electrons at birth and their residual momentum need to be further reduced. Intuitively, a second injection laser pulse with the same wavelength and much smaller spot size may appear to be the solution. However, simulations and analytical estimates show that for such a scheme the electrons will have large transverse residual momenta due to the ponderomotive acceleration ($\propto \nabla I$) induced by the small spot size of the injection laser. Here we describe results from simulation C where a circularly polarized 800 nm drive laser pulse with an $a_0 = 1.2$, which is high enough to drive a nonlinear wake but low enough as not to ionize the K-shell electrons of nitrogen, is propagating through a gas mixture. An injection pulse has the same wavelength, a focal spot of 2.5 μm, a pulse duration of 10 fs and an $a_0 = 2.0$, which is intense enough to ionize the K-shell electrons of nitrogen. With these parameters, the spot size of the injected electrons is reduced to $\sigma_{x0} = 0.71$ μm, but the residual momenta, $\sigma_{p0,x} = 0.77\,mc$, $\sigma_{p0,y} = 0.58\,mc$, are still quite high due to the ponderomotive acceleration, which leads to final emittances $\sim 1600\,\text{nm}\,(\hat{x})$ and $\sim 800\,\text{nm}\,(\hat{y})$.

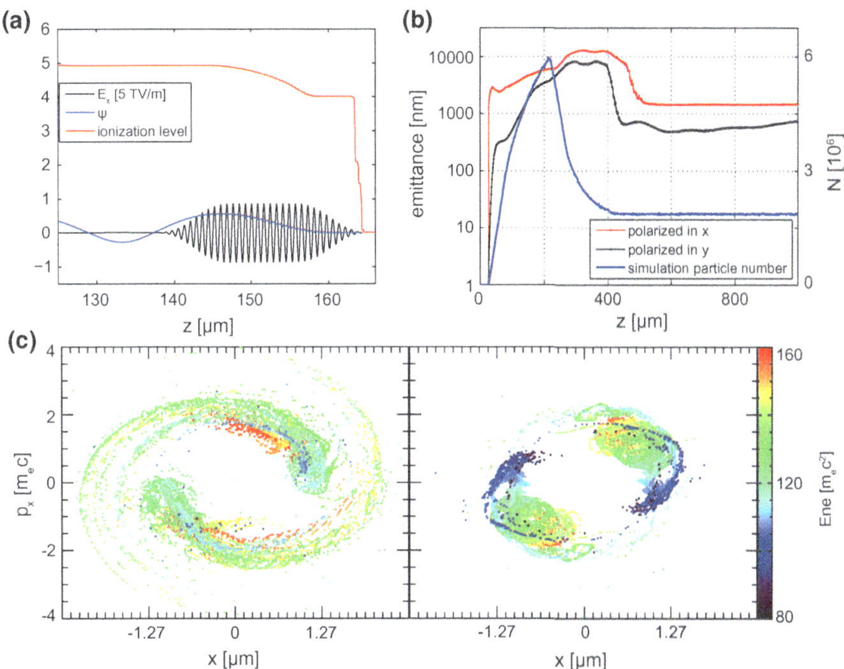

Fig. 2.22 2D OSIRIS simulation (B) of a 800 nm laser pulse with $a_0 = 1.4$ propagates through a mix of pre-ionized plasma ($n_e = 2.1 \times 10^{18}$ cm^{-3}) and C ($n_C = 0.01\, n_e$). **a** The laser electric field E_x, the normalized pseudo potential ψ and the ionization state of carbon atoms on axis. The laser longitudinal profile is a gaussian like 5th order symmetric polynomial $10t'^3 - 15t'^4 + 6t'^5$, where $t' = (t - t_0)/t_{rise}$ for the envelope rise and $t' = (t_0 + t_{flat} - t)/t_{fall}$ for the envelope fall. In this simulation $t_{rise} = t_{fall} = 34$ fs, $t_{flat} = 25$ fs. **b** The emittance evolution along (red) and perpendicular to the polarization direction (black), and the evolution of the particle number in the simulation (blue). The injected charge is about 20 pC by assuming a cylindrical distribution around z-axis. **c** The $x - p_x$ phase spaces at $z = 1$ mm for the laser polarized in (left) and out (right) of the simulation plane (Reprinted from Ref. [57]. Copyright with kind permission from American Physical Society)

To reduce the residual momenta of the injected electrons significantly, an injection laser with much shorter wavelength is needed, we call this the two color scheme. For such an injection laser, the vector potential a_0 that is needed for freeing the electrons through tunneling ionization is much reduced (due to the simple relation $a_0 \propto E_0\lambda$), therefore it can significantly reduce the residual momenta. As we noted in the above, this is inherently a multi-dimensional process. Next, we give an example from 2D PIC simulation by using an injection laser with $\lambda = 80$ nm (simulation D). Additional detail is given in Table 2.2 and in the figure capture to Fig. 2.23. As shown in Fig. 2.23a, the injection laser with $a_1 = 0.25$ is focused down to 0.64 μm to ionize the K-shell electrons of nitrogen in a very small transverse region. The transverse birth size is $\sigma_{x0} = 0.24\,\mu$m, and the residual momenta for the laser polarized in and out the simulation plane are $\sigma_{p0,x} = 0.066\,mc$, $\sigma_{p0,y} = 0.023\,mc$ respectively. The injection

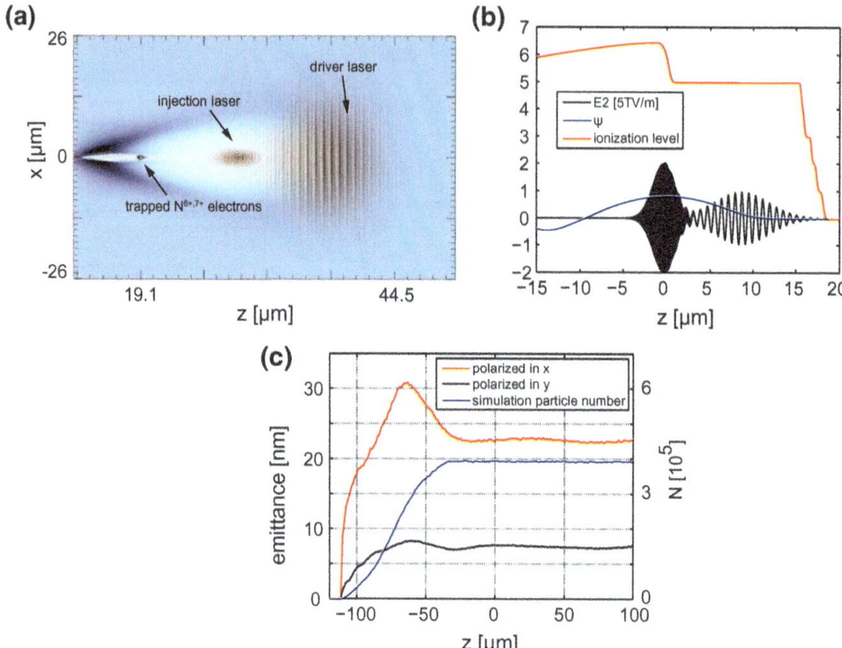

Fig. 2.23 2D OSIRIS simulation (D) of the two color ionization injection scheme using a 800 nm laser as the drive laser and a 80 nm laser for injection ($n_e = 1.7 \times 10^{18}$ cm^{-3}). The circularly polarized 800 nm laser pulse with spot size 10 μm and $a_0 = 1.2$ propagates to the right in a mixture of N^{5+} ($n_{N^{5+}} = 0.2 n_e$) and pre-ionized plasma. The linearly polarized 80 nm laser pulse with spot size 0.64 μm and $a_0 = 0.25$ is used to ionize and inject the electrons into the wake. The longitudinal profile of the driver laser and the injection laser are both gaussian with $\tau_{FWHM} = 18$ fs and 7 fs, respectively. $z = 0$ μm is the focal plane of the injection laser. **b** Snapshot of the charge density distribution of the wake electrons, the K-shell electrons of nitrogen, and the electric field in x direction. **b** The laser electric field E_x, the normalized pseudo potential ψ of the plasma wake and the ionization state of nitrogen atoms on axis. **c** The emittance evolution along (red) and perpendicular to the polarization direction (black), and the evolution of the particle number in the simulation (blue). The injected charge is about 1.5 pC by assuming a cylindrical distribution around z-axis (Reprinted from Ref. [57]. Copyright with kind permission from American Physical Society)

distance is about 77 μm $\approx 19 c/\omega_p$, which is long enough for transverse phase mixing and emittance saturation to occur. As shown in Fig. 2.23b, the emittances saturate at 20 nm($\hat{x}$) and 8 nm($\hat{y}$), which are significantly better than for the single wavelength case.

In the previous simulation, the injection laser power is about 85 GW and the total energy within the pulse is about 0.6 mJ which is sufficient to ionize the sixth electron of N. Currently such a powerful laser at 80 nm does not exist. For example extreme ultraviolet (EUV) sources based on high-order harmonics generation (HHG) typically have energies of a few μJ [58]. Therefore an experiment based on the above parameters will not be possible without improvement in the performance of such sources.

We therefore examine in 2D PIC simulations using a $10\,\mu$m laser pulse to drive the wake and a 400 nm laser pulse to inject electrons into this wake. Lasers with a longer wavelength have a larger vector potential for fixed laser intensity, therefore it is advantageous to use such a longer wavelength laser with large ponderomotive potential to drive nonlinear wakes and lower intensity to limit K-shell ionization. High power short pulse lasers needed for the injection pulse (with energy more than 10 mJ) can be obtained for wavelengths as short as 200 nm by frequency up-conversion of a 800 nm lasers. On the other hand, high power lasers with a mid-infrared wavelength to generate the wake are currently under development. For example, a 100 TW short pulse CO_2 ($10\,\mu$m wavelength) laser is currently under construction at Accelerator Test Facility of Brookhaven National Laboratory [59] while a 15 TW peak power CO_2 laser is currently operational, albeit with a 2 ps laser pulse at UCLA [60]. Such a scheme is therefore realizable experimentally with modest further improvements to existing technology.

In our simulation (simulation E), oxygen is used for the injection gas, where the first five electrons from oxygen are used to form the wake, and the sixth electron of oxygen atoms is used to supply the trapped electrons. We also use a preformed plasma. A circularly polarized laser ($\lambda_0 = 10$ μm, $a_0 = 1.4$, $E_{peak} = 0.45$ TV/m) drives a nonlinear wake in the blowout regime. A linearly polarized short wavelength laser ($\lambda_1 = 400$ nm, $a_1 = 0.09$, $E_{peak} = 0.72$ TV/m) trailing the driving laser is tightly focused to ionize the sixth electron with small transverse size and low residual momentum. In Fig. 2.24a 2D contour plot of the drive laser, injection laser, background electrons and trapped O^{5+} electrons are shown. Lineouts of the lasers, the wake potential and the ionization levels of the oxygen are shown in Fig. 2.24b. The power of the driver and the injection laser are 17 TW (6.4 J in 352 fs) and 22 GW (1 mJ in 44 fs), respectively.

The birth size and the residual momentum of the trapped electrons are $\sigma_{x0} = 1.2$ μm, $\sigma_{p0,x} = 1.8 \times 10^{-2}\,mc$, $\sigma_{p0,y} = 2.3 \times 10^{-3}\,mc$, respectively. The emittance evolution of the injected electron beam is shown in Fig. 2.24c, and one can see that the emittances saturate at 50 nm after the initial fast growth. The total injected charge is estimated as 4 pC by assuming the injected beam is cylindrically symmetric along z-axis. We note that due to the relatively long injection distance (0.45 mm $\approx 9c/\omega_p$), the slice energy spread of the injected beam reaches a few MeV, as shown in Fig. 2.25b. We note that similar simulation of two-color ionization injection with co-propagating geometry was also performed by L. L. Yu et al. at LBNL, with laser pulses of 5 μm and 400 nm wavelengths [26].

2.8.3 The Two-Color Ionization Injection: Transverse Injection

In the above longitudinal injection scheme, the intensity of the injection laser is large enough to release the trapped electrons when the pulse is near its focal plane. Before

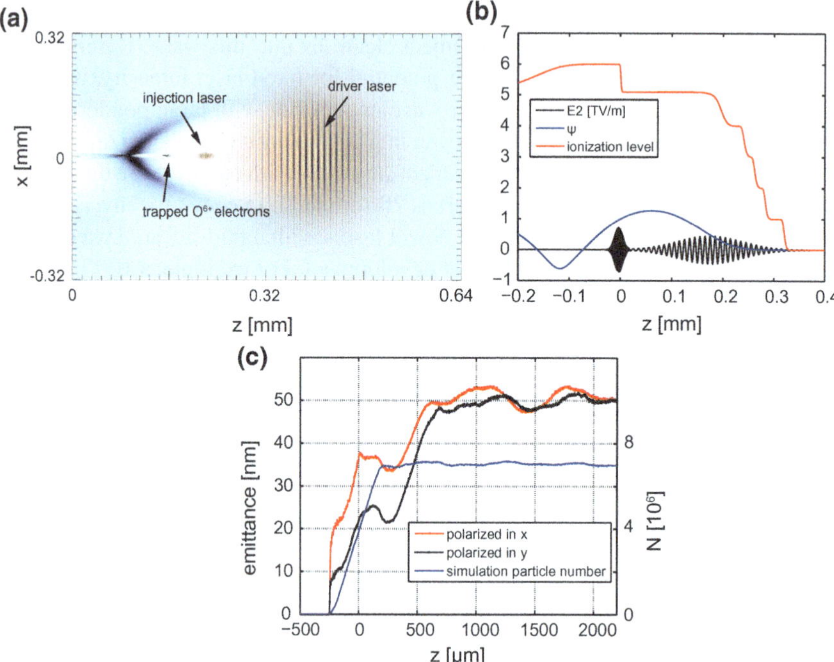

Fig. 2.24 **a** 2D OSIRIS simulation (E) of the two color ionization injection scheme using a $10\,\mu$m driving laser and a 400 nm injection laser. The density of the pre-ionized plasma electrons and the O^{5+} ions are $n_e = 1.12 \times 10^{16}$ cm^{-3} and $n_{O^{5+}} = 0.2n_e$ respectively. The driving laser has a longitudinal gaussian profile with $\tau_{FWHM} = 352$ fs, and with a transverse spot size $w_0 = 142\,\mu$m. The injection laser also has a longitudinal gaussian profile with $\tau_{FWHM} = 44$ fs, and with a transverse spot size $w_1 = 4.8\,\mu$m. $z = 0$ mm is the focal plane of the injection laser. **b** The laser electric field E_x, the normalized pseudo potential ψ of the plasma wake and the ionization state of oxygen atoms on axis. **c** The emittance evolution along (red) and perpendicular to the polarization direction (black), and the evolution of the particle number in the simulation (blue) (Reprinted from Ref. [57]. Copyright with kind permission from American Physical Society)

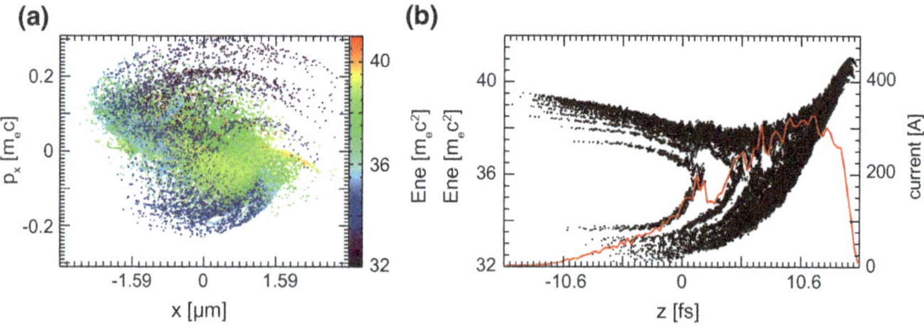

Fig. 2.25 Snapshots of the phase space from simulation (E) of Fig. 5. **a** The transverse phase space at $z = 2.2$ mm; **b** the longitudinal phase space and the current profile at $z = 2.2$ mm (Reprinted from Ref. [57]. Copyright with kind permission from American Physical Society)

and after the focal plane, the intensity of the laser pulse decreases and the ionization ceases. So the injection distance is on the order of the Rayleigh length of the injection laser which is about several or dozens of plasma skin depths for typical parameters. The long injection distance leads to the saturation of the emittance which is much larger than the thermal emittance. And the energy spread in each slice of the injected beam grows with the injection distance, and typically reaches a few MeV by the end of the injection process. To suppress the transverse phase mixing and the significantly reduce the slice energy spread, we propose to use two transverse colliding pulses to limit the injection distance. This has also been shown to reduce the emittance and slice energy spread in a beam driver case [23].

Two counter-propagating short wavelength laser pulses ($\lambda = 400$ nm) moving along the $+$ and $- \hat{x}$-axis directions are synchronized with the driver laser so that they overlap inside the nonlinear wake near the point where the longitudinal electric field vanishes. In Ref. [23] where a particle beam driver was used the ionizing lasers had an a_0 of 0.016 as wavelength of 800 nm, and the ionizing gas was He. Because we are using a laser driver which has a larger intensity, oxygen is used and the ionizing lasers have a normalized potential $a_0 = 0.06$ (not sufficient to ionize the sixth L-shell electron of oxygen). The laser intensity exceeds the ionization threshold only where the two lasers overlap, and a large fraction of O^{5+} ions within this volume is now ionized to O^{6+} as shown in Fig. 2.26a. As the lasers travel past the collision point, the injection ceases. These laser-ionized oxygen electrons then respond to the wake fields and are rapidly accelerated to a longitudinal velocity close to c as they slip backwards to the rear of the nonlinear wake. They then began to move nearly synchronously with the wake, as depicted in Fig. 2.26b. Figure 2.26c shows the emittance evolution in this simulation, and one can see that emittance of $\sim$60 nm is achieved. Figure 2.26d shows the longitudinal phase space of the injected beam, which confirms clearly that the slice energy spread (as small as 30 keV) is much reduced comparing with the longitudinal injection case (a few MeV). Such low slice energy spread may have significant advantages in applications relevant to future coherent light sources driven by plasma accelerators.

The peak electric field of the injection lasers is $E_0 = 0.48$ TV/m and the corresponding pulse duration threshold is $\tau_{crit} \approx 5$ fs, therefore the pulse duration used in the simulation τ_{FWHM} is larger than the threshold, leading a residual momentum in x direction which is on the order of a_0. The birth size and the residual momentum of the trapped electrons from the simulation are $\sigma_{x0} = 1.1$ μm, $\sigma_{p0.x} = 0.04\, mc$, leading to a thermal emittance $\epsilon_{th} = 44$ nm. One can see the emittance of the injected beam is close to the thermal emittance, which means the transverse phase mixing is ideally suppressed in this transverse injection scheme.

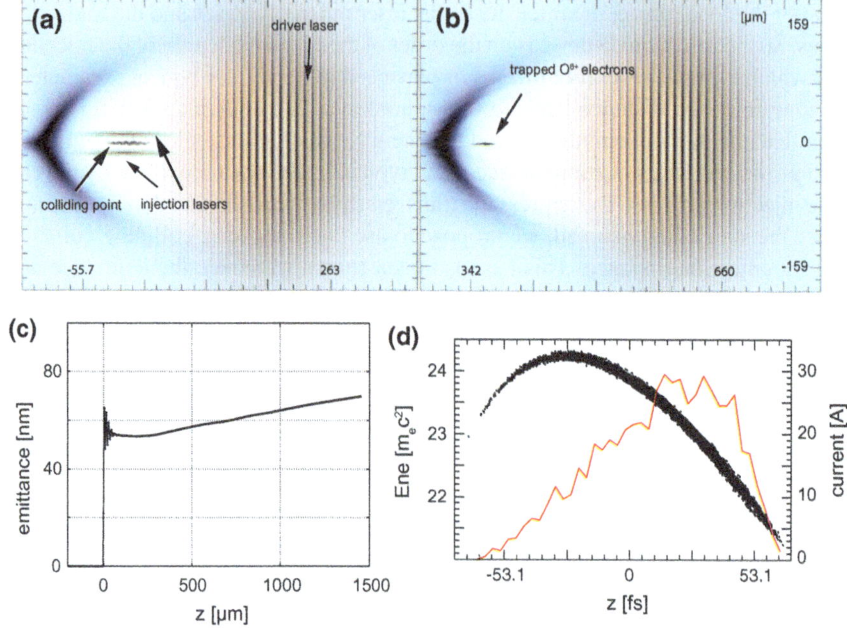

Fig. 2.26 OSIRIS simulations of the two color ionization injection scheme by two transverse colliding pulses ($n_e = 1.12 \times 10^{16}$ cm^{-3}). The circularly polarized $10\,\mu$m laser pulse with spot size $142\,\mu$m, $\tau_{FWHM} = 352$ fs and $a_0 = 1.4$ propagates to the right in a mixture of O^{5+} ($n_{O^{5+}} = 0.2n_e$) and a pre-ionized plasma. The two linearly polarized 400 nm laser pulse with spot size $48\,\mu$m and $a_0 = 0.06$ ionize the O^{5+} to generate the trapped electrons. $z = 0$ mm is the plane where the two injection laser pulses—spatially and temporally overlap. The two injection laser pulses have a gaussian longitudinal profile with $\tau_{FWHM} = 13$ fs. **a** Snapshot of the charge density distribution of wake electrons, the K-shell electrons of nitrogen, and the electric field in x direction and in z direction at the laser pulses's collision time, **b** 1 ps after the collision when the injected electrons become trapped in the wake. **c** The emittance evolution in x direction. The injected charge is about 2 pC by assuming a cylindrical distribution along x-axis. **d** The $z - p_z$ phase space and the current profile of the injected beam at $z = 1.5\,\mu$m (Reprinted from Ref. [57]. Copyright with kind permission from American Physical Society)

2.9 Intrinsic Phase Space Discretization in Laser Triggered Ionization Injection

As shown in Sect. 2.2, the ionization probability in the tunneling limit has a strong dependence on the phase of the injection laser $\theta = \omega_0(t - z/c)$, i.e, the electrons are mostly released near the peak of the laser electric field, where ω_0 is the laser frequency. This phase dependent ionization leads to an intrinsic phase space discretization in optical frequencies when the electrons are freed and this discretization pattern is mapped to the final phase space when the electrons are boosted to relativistic energy based on the trapping condition. This phase space discretization is ubiquitous in all kinds of laser triggered ionization injection schemes. In this section, we will discuss these interesting phenomena and the potential applications in details.

2.9.1 Single Laser Pulse Case

First we consider a single laser driven ionization injection as shown in Fig. 2.27a. An 800 nm laser pulse with normalized vector potential $a_0 = 2$ propagates into a mixture of pre-ionized plasma and nitrogen gas. The wake excitation and injection of the K-shell electrons of nitrogen are examined using the three-dimensional (3D) particle-in-cell (PIC) code OSIRIS in cartesian coordinates using a moving window. The simulation window has a dimension of $38.1 \times 63.5 \times 63.5$ μm with $1500 \times 500 \times 500$ cells in the $\hat{x}$, $\hat{y}$ and $\hat{z}$ directions, respectively. This corresponds to cell sizes of k_0^{-1} in the $\hat{x}$ and $\hat{y}$ directions and $0.2\ k_0^{-1}$ in the $\hat{z}$ direction, where k_0 is the wavenumber of the laser pulse.

The (ξ, x) space distribution of the trapped electrons when they are ionized is shown in Fig. 2.27b. Due to the phase dependent ionization probability, the initial electron distribution has a strong modulation on $2\omega_0$. After released, the electrons slip back to the tail of the wake and are accelerated by the longitudinal electric field in the wake. If its velocity equals the phase velocity of the wake, the electron is defined as being trapped. The phase space after the electrons are boosted to relativistic energy can be mapped from the initial phase space based on the trapping condition. The (ξ, x, p_z) phase space at $z = 0.5$ mm is shown in Fig. 2.27c and the sliced structures are clearly seen. The current profile and the bunching factor $b(k)$ of the tapped electrons are shown in Fig. 2.27d, where $b(k) = \int dz f(z)\exp(ikz)$ and $f(z)$ is the normalized distribution of the trapped electrons.

When the injection distance $L_{inj} = 90$ μm $\approx 24c/\omega_p$ which is the range of nitrogen distribution in Fig. 2.27a–d, a strong modulation at $4\omega_0$ can be seen in the current profile and the bunching factor plot as shown in Fig. 2.27d. If the injection distance is increased to $L_{inj} = 190$ μm $\approx 50c/\omega_p$, the modulation in the current profile is hidden in Fig. 2.27f but the sliced structure is still clearly seen in the (ξ, x, p_z) phase space as shown in Fig. 2.27e.

2.9.2 Beam Driver with a Laser Injector

By using two pulses to separate the wake excitation process and the injection process, the initial and final phase space of the trapped electrons can be more controllable, and more clear bunched structure can be found in the current profile. In a relativistic beam driver case, the phase velocity of the nonlinear wake is equal to the velocity of the driver beam. The dephasing length which is defined as the distance the trapped electron outruns the acceleration phase is proportional to γ_p^2. If the energy of the driver beam is high, e.g., hundreds of MeV, the trapped electrons are lon-

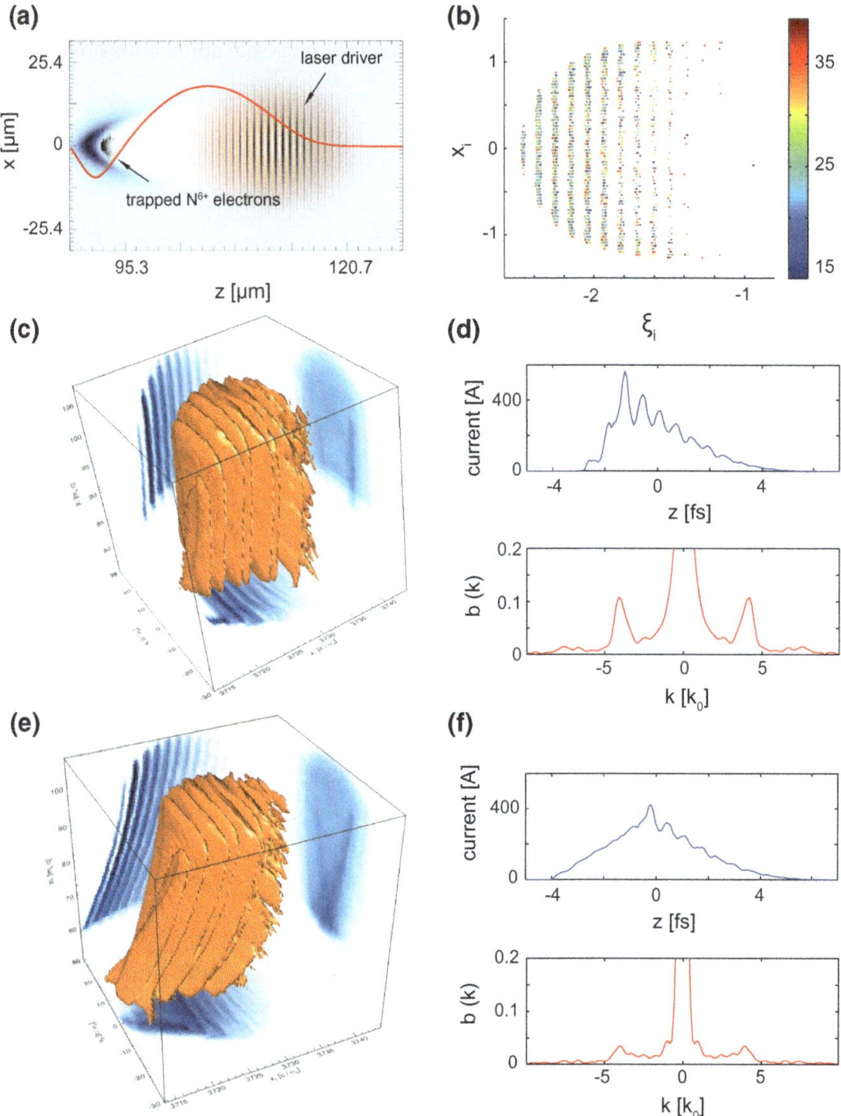

Fig. 2.27 An 800 nm laser pulse with $a_0 = 2$, $w_0 = 14$ μm and $\tau_{FWHM} = 26$ fs propagates to the right in a mixture of pre-ionized plasma and nitrogen gas ($n_p = 2 \times 10^{18}$ cm^{-3}). The nitrogen is distributed from $z = 0.038$ mm to $z = 0.127$ mm in **c** and **d**, and from $z = 0.038$ mm to $z = 0.228$ mm in **e** and **f**. The laser is focused at $z = 0$ mm plane. **a** Snapshot of the charge density distribution of the background electrons, the K-shell electrons of nitrogen, and the electric field in x direction. The red line is the pseudo potential in the center with $x = y = 0$ μm. **b** The (ξ, x) space distribution when the electrons are ionized. The color represents the time when the electrons are ionized. The (ξ, x, p_z) phase space **c**, the current profile and the bunching factor **d** at $z = 0.5$ mm when the nitrogen has a 90 μm range and $n_N = 0.01 n_p$. The (ξ, x, p_z) phase space (**e**), the current profile and the bunching factor (**f**) at $z = 0.5$ mm when the nitrogen has a 190 μm range and $n_N = 0.005 n_p$

gitudinally frozen in the wake after the initial slippage. This longitudinal position can be defined as the 'final' position of the electron. By applying the trapping condition, the final relative longitudinal position of the injected electron can be obtained as $\xi_f \approx \sqrt{4 + \xi_i^2 + r_i^2 - r_f^2}$, where the maximum radius of the wake $r_m > 2$ is assumed.

In the wake, the electron conducts transverse betatron oscillation under the focusing and acceleration fields with $r \sim r_i \gamma^{-1/4}$. Therefore, if the injection laser is focused to a small spot size with $w_0 \ll 1$, the final relative longitudinal position of the trapped electron has a simple dependence on the initial relative longitudinal position as $\xi_f \approx \sqrt{4 + \xi_i^2}$. The density modulation of the ξ_i can be mapped to the ξ_f, thus an electron bunch train can be generated. When $\delta\xi_i, r_i \ll 1$, the final position can be approximated as

$$\xi_f \approx \sqrt{4 + \bar{\xi}_i^2} + \frac{\delta\xi_i \bar{\xi}_i}{\sqrt{4 + \bar{\xi}_i^2}} + \frac{(\delta\xi_i)^2 + r_i^2}{2\sqrt{4 + \bar{\xi}_i^2}} \tag{2.54}$$

where $\bar{\xi}_i$ is the mean value of ξ_i and $\delta\xi_i \equiv \xi_i - \bar{\xi}_i$. If we define the laser frequency as the fundamental frequency, it is straightforward to obtain the harmonic number of the modulation frequency in the current profile as

$$n = 2\frac{\sqrt{4 + \bar{\xi}_i^2}}{|\bar{\xi}_i|} \tag{2.55}$$

where the factor 2 is from the ionization process.

In order to get an effective modulation of the final current profile, the contributions from r_i and $\delta\xi_i$ to the final position ξ_f need be small than half wavelength of the final modulation, i.e.,

$$(\delta\xi_i)^2 + r_i^2 < \lambda_{inj}\bar{\xi}_i/2 \tag{2.56}$$

Here we discuss a beam driven and laser injection case. A relativistic electron beam propagates into a mixture of pre-ionized plasma and He^{1+} ions. The electric field of the electron beam is low enough to keep the 2nd electron of helium bounded and high enough to drive a highly nonlinear wake. An 800 nm injection laser is focused into the wake with $w_0 = 2$ μm as shown in Fig. 2.28a. The current profile of the electron beam at $z = 0.1$ mm and the bunching factor $b(k)$ are shown in Fig. 2.28d. One can see the bunching factor peaks at $\sim 3k_0$ which has a good agreement with Eq. 2.55. The simulation window has a dimension of $127 \times 127 \times 127$ μm with $5000 \times 1000 \times 1000$ cells in the $\hat{x}$, $\hat{y}$ and $\hat{z}$ directions, respectively. This corresponds to cell sizes of k_0^{-1} in the $\hat{x}$ and $\hat{y}$ directions and $0.2\ k_0^{-1}$ in the $\hat{z}$ direction, where k_0 is the wavenumber of the injection laser pulse. There are 4 particles per cell to represent the He^{1+} ions.

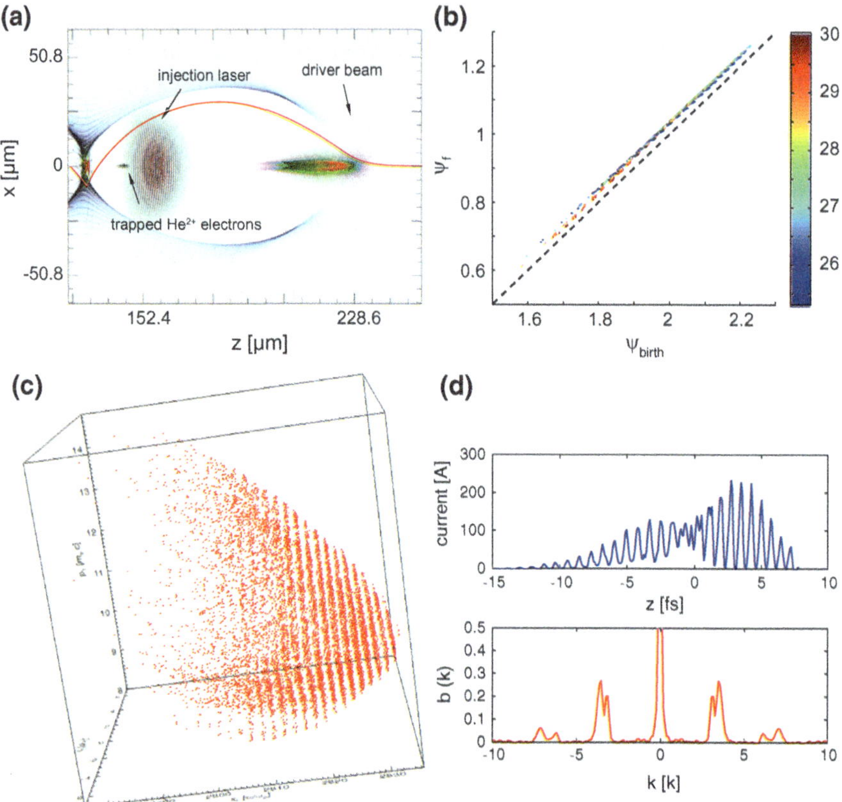

Fig. 2.28 An electron beam and a following laser pulse propagate to the right in a mixture of pre-ionized plasma and He^{1+} ions ($n_p = 1.74 \times 10^{17}$ cm^{-3}, $n_{He^{1+}} = 0.1 n_p$). Driver beam: $E_b = 100$ MeV, $\sigma_r = 8.9$ μm, $\sigma_z = 10.6$ μm, $I_b = 19$ kA; Injection laser: $\lambda = 800$ nm, $a_0 = 0.12$, $w_0 = 2$ μm, $\tau_{FWHM} = 32$ fs. The laser is focused at $z = 0$ mm. **a** Snapshot of the charge density distribution of the background electrons, the 2rd electron of helium, and the laser electric field. The red line is the pseudo potential in the center with $x = y = 0$ μm. **b** The dependence of the final ψ_f on the ψ_{birth} when the electrons are ionized. The color represents the ionization time. The black dashed line represents $\psi_f - \psi_{birth} = -1$. **c** The (ξ, x, p_z) phase space distribution of the trapped electrons at $z = 0.1$ mm. **d** The current profile and the bunching factor of the trapped electrons at $z = 0.1$ mm with $\langle \gamma_b \rangle = 10.2$, $\sigma_\gamma = 1.32$

By replacing the 800 nm injection laser by a 200 nm injection laser, an electron bunch with strong UV frequency modulation can be generated. A 200 nm injection laser with $a_0 = 0.023$, $w_0 = 2$ μm are used to release the 2nd electron of helium. The peak power of the total energy of the injection laser is $P_{inj} = 1.14$ GW, $E_{inj} = 0.03$ mJ. The current profile and the bunching factor at $z = 0.1$ mm are shown in Fig. 2.29a, b. The rms length of the injected beam is $\tau_{rms} = 3$ fs, and the total injected charge is $Q = 3.9$ pC. In this simulation, an external field model is used to play the role of the nonlinear wake for saving the simulation cpu hours.

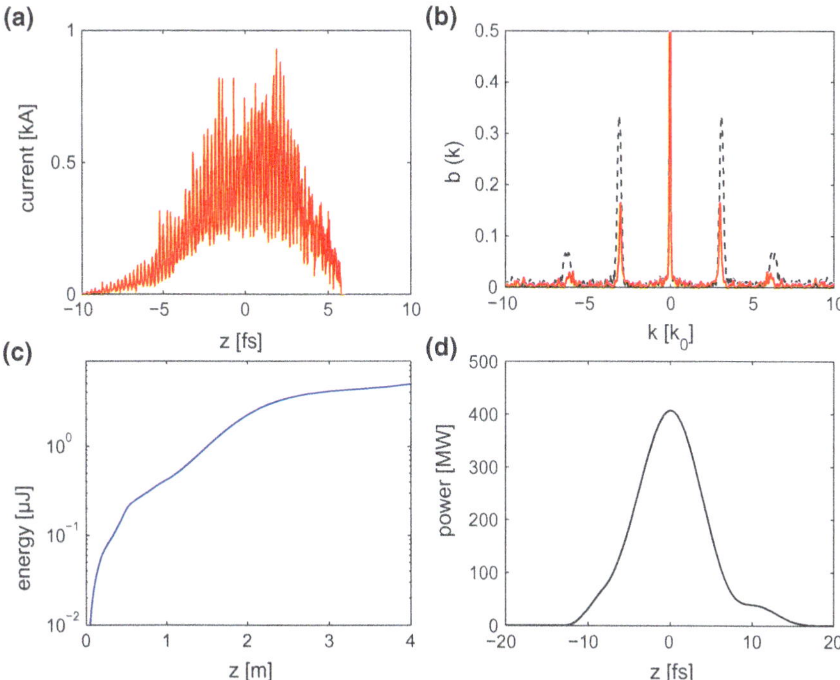

Fig. 2.29 The current profile (**a**) and the bunching factor **b** of the trapped electrons by using 200 nm injection laser at $z = 0.1$ mm. The red lines are the results when $n_{\mathrm{He}^{1+}} = 0.3\,n_p$ and the black dashed line is the result when $n_{\mathrm{He}^{1+}} = 3 \times 10^{-5} n_p$. **c** The pulse energy along the undulator; **d** the output radiation pulse profile at $z = 3$ m

Due to the small spot size and low intensity of the injection laser, the emittance and energy spread of the trapped beam are both very small, $\epsilon_x = 10.9$ nm, $\epsilon_y = 10.6$ nm, $\sigma_\gamma = 3.2$. If the energy of the electron beam can be boosted to ~ 500 MeV, such a beam can potentially generate coherent EUV radiation at the bunch spacing frequency by going through a short undulator. For example, the electron beam with $\langle \gamma_b \rangle = 1068.9$ and $\sigma_\gamma = 3.2$ propagates into a planar undulator with wavelength $\lambda_u = 3$ cm and normalized undulator parameter $K \equiv (e/2\pi mc)B_p\lambda_u = 2$, where B_p is the on-axis peak magnetic field of the undulator. The undulator is resonant at the modulation wavelength of the electron beam, $\lambda_r = 65.6$ nm. The transverse beam sizes are optimized at $\sigma_{x,y} = 8$ μm. The output radiation power saturates at $P_{sat} = 400$ MW in 3 m undulator as shown in Fig. 2.29c, d.

2.10 Summary

In this chapter, we studied the evolution of beam phase space in ionization injection in plasma wakefields using theory and simulations. Several important physical processes in ionization injection have been examined systematically. Two key phase mixing processes, namely longitudinal and transverse phase mixing, are found to be responsible for the complex emittance dynamics that shows initial rapid growth followed by oscillation, decay, and eventually slow growth to saturation. An analytic theory is developed to include the effects of injection distance (time), acceleration distance, wakefield structure, and nonlinear space charge forces. Formulas for the emittance in the low and high charge regimes are presented and verified through PIC simulations with good agreements. A two-color ionization injection scheme for generating high quality beams has been examined through PIC simulations. In the end, the intrinsic phase space discretization in laser triggered ionization injection is analyzed.

References

1. Esarey E, Schroeder CB, Leemans WP (2009) Physics of laser-driven plasma-based electron accelerators. Rev Mod Phys 81:1229–1285
2. Leemans WP, Nagler B, Gonsalves AJ et al (2006) GeV electron beams from a centimetre-scale accelerator. Nat Phys 2(10):696–699
3. Hafz NAM, Jeong TM, Choi IW et al (2008) Stable generation of GeV-class electron beams from self-guided laser-plasma channels. Nat Photon 2(9):571–577
4. Kneip S, Nagel SR, Martins SF et al (2009) Near-GeV acceleration of electrons by a nonlinear plasma wave driven by a self-guided laser pulse. Phys Rev Lett 103:035002
5. Clayton CE, Ralph JE, Albert F et al (2010) Self-guided laser wakefield acceleration beyond 1 GeV using ionization-induced injection. Phys Rev Lett 105:105003
6. Liu JS, Xia CQ, Wang WT et al (2011) All-optical cascaded laser wakefield accelerator using ionization-induced injection. Phys Rev Lett 107:035001
7. Pollock BB et al (2011) Demonstration of a narrow energy spread, $\sim$ 0.5 GeV electron beam from a two-stage laser wakefield accelerator. Phys Rev Lett 107:045001
8. Wang X, Zgadzaj R, Fazel N et al (1988) Quasi-monoenergetic laser-plasma acceleration of electrons to 2 GeV. Nat Commun 2013:4
9. Kim HT, Pae KH, Cha HJ et al (2013) Enhancement of electron energy to the multi-GeV regime by a dual-stage laser-wakefield accelerator pumped by petawatt laser pulses. Phys Rev Lett 111:165002
10. Leemans WP, Gonsalves AJ, Mao HS et al (2014) Multi-GeV electron beams from capillary-discharge-guided subpetawatt laser pulses in the self-trapping regime. Phys Rev Lett 113:245002
11. Muggli P, Blue BE, Clayton CE et al (2004) Meter-scale plasma-wakefield accelerator driven by a matched electron beam. Phys Rev Lett 93:014802
12. Hogan MJ, Barnes CD, Clayton CE et al (2005) Multi-GeV energy gain in a plasma-wakefield accelerator. Phys Rev Lett 95:054802
13. Blumenfeld I, Clayton CE, Decker FJ et al (2007) Energy doubling of 42 GeV electrons in a metre-scale plasma wakefield accelerator. Nature 445(7129):741–744
14. Geddes CGR, Nakamura K, Plateau GR et al (2008) Plasma-density-gradient injection of low absolute-momentum-spread electron bunches. Phys Rev Lett 100:215004

15. Gonsalves A, Nakamura K, Lin C et al (2011) Tunable laser plasma accelerator based on longitudinal density tailoring. Nat Phys 7(11):862–866
16. Faure J, Rechatin C, Norlin A et al (2006) Controlled injection and acceleration of electrons in plasma wakefields by colliding laser pulses. Nature 444(7120):737–739
17. Rechatin C, Faure J, Ben-Ismail A et al (2009) Controlling the phase-space volume of injected electrons in a laser-plasma accelerator. Phys Rev Lett 102:164801
18. Kotaki H, Daito I, Kando M et al (2009) Electron optical injection with head-on and counter-crossing colliding laser pulses. Phys Rev Lett 103:194803
19. Oz E, Deng S, Katsouleas T et al (2007) Ionization-induced electron trapping in ultrarelativistic plasma wakes. Phys Rev Lett 98:084801
20. Pak A, Marsh KA, Martins SF et al (2010) Injection and trapping of tunnel-ionized electrons into laser-produced wakes. Phys Rev Lett 104:025003
21. McGuffey C, Thomas AGR, Schumaker W et al (2010) Ionization induced trapping in a laser wakefield accelerator. Phys Rev Lett 104:025004
22. Hidding B, Pretzler G, Rosenzweig JB et al (2012) Ultracold electron bunch generation via plasma photocathode emission and acceleration in a beam-driven plasma blowout. Phys Rev Lett 108:035001
23. Li F, Hua JF, Xu XL et al (2013) Generating high-brightness electron beams via ionization injection by transverse colliding lasers in a plasma-wakefield accelerator. Phys Rev Lett 111:015003
24. Bourgeois N, Cowley J, Hooker SM (2013) Two-pulse ionization injection into quasilinear laser wakefields. Phys Rev Lett 111:155004
25. Ossa A, Grebenyuk J, Mehrling T et al (2013) High-quality electron beams from beam-driven plasma accelerators by wakefield-induced ionization injection. Phys Rev Lett 111:245003
26. Yu LL, Esarey E, Schroeder CB et al (2014) Two-color laser-ionization injection. Phys Rev Lett 112:125001
27. Kirby N, Blumenfeld I, Clayton CE et al (2009) Transverse emittance and current of multi-GeV trapped electrons in a plasma wakefield accelerator. Phys Rev ST Accel Beams 12:051302
28. Chen M, Esarey E, Schroeder CB et al (2012) Theory of ionization-induced trapping in laser-plasma accelerators. Phys Plasmas 19(3)
29. Xi Y, Hidding B, Bruhwiler D et al (2013) Hybrid modeling of relativistic underdense plasma photocathode injectors. Phys Rev ST Accel Beams 16:031303
30. Schroeder CB, Vay JL, Esarey E et al (2014) Thermal emittance from ionization-induced trapping in plasma accelerators. Phys Rev ST Accel Beams 17:101301
31. Lu W (2006) Nonlinear plasma wakefield theory and optimum scaling for laser wakefield acceleration in the blowout regime. PhD dissertation UCLA
32. Tzoufras M, Lu W, Tsung FS et al (2008) Beam loading in the nonlinear regime of plasma-based acceleration. Phys Rev Lett 101:145002
33. Tzoufras M, Lu W, Tsung FS et al (2009) Beam loading by electrons in nonlinear plasma wakes. Phys Plasmas 16(5):056705
34. Marsh KA, Clayton CE, Joshi C et al (2011) Laser wakefield acceleration beyond 1 GeV using ionization induced injection. LLNLPROC-476791
35. Zeng M, Chen M, Sheng ZM et al (2014) Self-truncated ionization injection and consequent monoenergetic electron bunches in laser wakefield acceleration. Phys Plasmas 21(3)
36. Wang S, Clayton CE, Blue BE et al (2002) X-Ray emission from betatron motion in a plasma wiggler. Phys Rev Lett 88:135004
37. Lu W (2011) Self and controlled injection in multi-dimensional wakefields driven by lasers or charged beams. LPAW11, 20–24 June, Wuzhen, China
38. Delone NB, Krainov VP (1998) Tunneling and barrier-suppression ionization of atoms and ions in a laser radiation field. Physics-Uspekhi 41:469–85
39. Popov VS (2004) Tunnel and multiphoton ionization of atoms and ions in a strong laser field (Keldysh theory). Physics-Uspekhi 47(9):855–885
40. Keldysh L (1964) Tunneling theory of multi-photon absorption. Zh Eksp Teor Fiz 1945:47
41. Ammosov MV, Delone NB, Krainov VP (1986) JETP. Sov Phys 64:1191

42. Yudin GL, Ivanov MY (2001) Nonadiabatic tunnel ionization: looking inside a laser cycle. Phys Rev A 64:013409
43. Delone NB, Krainov VP (1991) Energy and angular electron spectra for the tunnel ionization of atoms by strong low-frequency radiation. J Opt Soc Am B 8(6):1207–1211
44. Quesnel B, Mora P (1998) Theory and simulation of the interaction of ultraintense laser pulses with electrons in vacuum. Phys Rev E 58(3, B):3719–3732
45. Mendonca JT (1983) Threshold for electron heating by two electromagnetic waves. Phys Rev A 28:3592–3598
46. Vehn J, Sheng Z (1999) On electron acceleration by intense laser pulses in the presence of a stochastic field. Phys Plasmas 6(3):641–644
47. Sheng ZM, Mima K, Sentoku Y et al (2002) Stochastic Heating and Acceleration of Electrons in Colliding Laser Fields in Plasma. Phys Rev Lett 88:055004
48. Sheng ZM, Mima K, Zhang J et al (2004) Efficient acceleration of electrons with counterpropagating intense laser pulses in vacuum and underdense plasma. Phys Rev E 69:016407
49. Moore CI, Ting A, McNaught SJ et al (1999) A laser-accelerator injector based on laser ionization and ponderomotive acceleration of electrons. Phys Rev Lett 82:1688–1691
50. Kaw PK, Kulsrud R (1973) Relativistic acceleration of charged particles by superintense laser beams. Phys Fluids 16:321
51. Corkum PB, Burnett NH, Brunel F (1989) Above-threshold ionization in the long-wavelength limit. Phys Rev Lett 62:1259–1262
52. Dowell DH, Schmerge JF (2009) Quantum efficiency and thermal emittance of metal photocathodes. Phys Rev ST Accel Beams 12(7):074201
53. Lu W, Huang C, Zhou M et al (2006) Nonlinear theory for relativistic plasma wakefields in the blowout regime. Phys Rev Lett 96:165002
54. Lu W, Huang C, Zhou M et al (2006) A nonlinear theory for multidimensional relativistic plasma wave wakefields. Phys Plasma 13:056709
55. Xu XL, Hua JF, Li F et al (2014) Phase-Space Dynamics of Ionization Injection in Plasma-Based Accelerators. Phys Rev Lett 112(3):035003. https://journals.aps.org/prl/abstract/10.1103/PhysRevLett.112.035003
56. Serafini L, Rosenzweig JB (1997) Envelope analysis of intense relativistic quasilaminar beams in rf photoinjectors: mA theory of emittance compensation. Phys Rev E 55:7565–7590
57. Xu XL, Wu YP, Zhang CJ et al (2014) Low emittance electron beam generation from a laser wakefield accelerator using two laser pulses with different wavelengths. Phys Rev ST Accel Beams 17(6):061301. https://journals.aps.org/prab/abstract/10.1103/PhysRevSTAB.17.061301
58. Lambert G et al (2005) Seeding the FEL of the SCSS Phase 1 facility with the 13th laser harmonic of a Ti: Sa laser produced in Xe gas. Proc Proc FEL 2004:155–158
59. Polyanskiy MN, Babzien M, Pogorelsky I et al (2013) Ultrashort-pulse CO_2 lasers: ready for the race to petawatt? In: Proceedings of XIX international symposium on high-power laser systems and applications. International Society for Optics and Photonics 86770G–86770G
60. Haberberger D, Tochitsky S, Joshi C (2010) Fifteen terawatt picosecond CO_2 laser system. Opt Express 18(17):17865–17875

Chapter 3
Coherent Phase Space Matching Using Longitudinally Tailored Plasma Structure

3.1 Introduction

One critical issue of plasma accelerators (PAs) is about staging, namely connecting stages of PAs or connecting PA with the traditional accelerator components, as illustrated in Fig. 3.1 with four typical scenarios. To see the necessity or sometimes the advantage of staging, here we analyze for four different applications: (1) staging PAs for TeV class collider, (2) staging PA and light source components such as undulators or storage rings, (3) staging closely packed PAs, and (4) staging RF-based injector and PA.

In case (1), for the collider concepts based on PAs recently developed [1, 2], each stage of PAs has its own driver, and each stage will boost the energy of the witness beam by 10s of GeV. The distance between stages needs to be larger than a meter to place beam transport components for coupling the fresh driver to the next stage. In case (2), for PAs driven light sources [3–6], the high quality electron beam needs to be coupled from the plasma wakes to the traditional accelerator components as shown in Fig. 3.1c. In case (3), for GeV-level electron beam generation, two-stage PAs have been demonstrated to generate electron beams with high stability and tunability [7–10]. In principle, by injecting an ultrashort low energy electron beam from a high density ($\sim 10^{19}\,\mathrm{cm}^{-3}$) plasma injector into a low density PA ($\sim 10^{17}\,\mathrm{cm}^{-3}$), the occupied acceleration phase by the beam is much reduced, therefore a much reduced relative energy spread can be achieved as the energy is boosted. In case (4), external injection schemes utilizing high quality electron beams from RF-based accelerators may potentially lead to stable, high quality acceleration [11, 12].

In all the cases above, as the beam propagating between different stages, seriously phase space mismatch may induce catastrophic emittance growth if a finite energy spread is present. Therefore beam quality conserving matching schemes between stages is highly needed for the future development of plasma acceleration. In this chapter we propose using longitudinally tailored plasma structure to match the beam between stages to conserve the beam quality. In Sect. 3.2 the mechanism of emittance growth between stages are discussed in details. In Sect. 3.3 we theoretically analyze

© Springer Nature Singapore Pte Ltd. 2020
X. Xu, *Phase Space Dynamics in Plasma Based Wakefield Acceleration*,
Springer Theses, https://doi.org/10.1007/978-981-15-2381-6_3

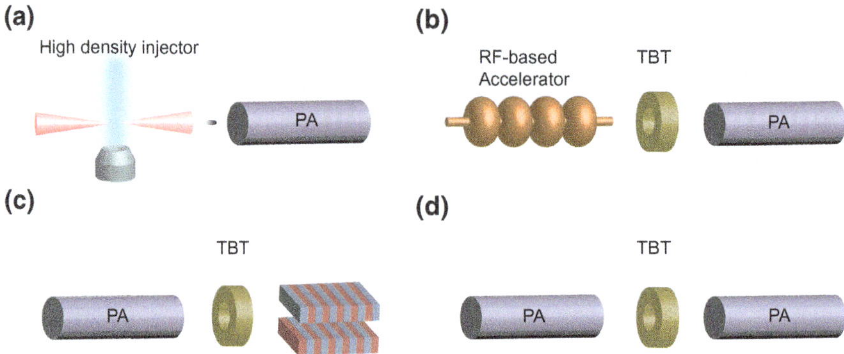

Fig. 3.1 Configurations of different stages. **a** Plasma based injector and PA; **b** RF-based injector and PA; **c** PA and undulator; **d** PAs. The yellow quadrupole represents the traditional beam transport (TBT)

how to use a longitudinally tailored plasma structure to match the beam between stages. In Sect. 3.4 PIC simulations are used to verify our theoretical design for different staging scenarios. A summary is given in Sect. 3.5.

3.2   The Emittance Growth Between Stages

In typical staging scenarios (Fig. 3.1), the beam needs to be coupled between stages with drastically different field strengths. In traditional accelerator, solenoids and quadrupoles are combined to guide the particles between stages. However, due to ultra high focusing gradient in the plasma waves ($G \equiv F_r/ecr = n_p e/2c\epsilon_0$, e.g., $G = 3 \times 10^6$ T/m when $n_p = 10^{17}$ cm^{-3}), the typical quadrupoles with gradients on the order of 10^3 T/m [13, 14] are not strong enough to confine the transverse motion between stages. As a result, beams experience orders of magnitude transverse size variation when propagating between the PAs and the traditional components, and their transverse motion becomes very sensitive to the energy spread: particles with different energy conduct transverse betatron oscillations with different betatron phases, leading to a catastrophic emittance growth [15–17].

To show how this emittance growth can occur, we start with an analysis of emittance. The transverse normalized emittance is defined as $\epsilon_n = \frac{1}{mc}\sqrt{\langle x^2 \rangle \langle p_x^2 \rangle - \langle x p_x \rangle^2}$, where $\langle \rangle$ represents an ensemble average over the beam distribution, x is the transverse position and p_x is the transverse momentum, m the electron rest mass and c the speed of light. To describe the profile of the phase space distribution, Courant-Snyder parameters are introduced as $\beta = \langle x^2 \rangle / \epsilon_{geo}, \alpha = \langle x x' \rangle / \epsilon_{geo}, \gamma = \langle x'^2 \rangle / \epsilon_{geo}$ which satisfy $\beta\gamma = 1 + \alpha^2$, where $x' = p_x/p_z$ is the particle slope, and $\epsilon_{geo} = \sqrt{\langle x^2 \rangle \langle x'^2 \rangle - \langle x x' \rangle^2}$ is the geometric emittance. The beta function is a measure of the beam size, alpha function represents the correlation between x and x', and gamma function is a measure of the spread in the particle slopes. In typical cases, the C-S parameters of an electron beam in the plasma wakefield accelerators are determined by the field structure of the nonlinear wake as $\beta = \sqrt{2 \langle \gamma_b \rangle} k_p^{-1}$, $\alpha = 0$, where $\langle \gamma_b \rangle$ is the mean value of the relativistic factor of the beam. For an electron beam with finite energy spread, the evolution of the normalized emittance evolution can be derived as

$$
\begin{aligned}
\epsilon_n &= \frac{1}{mc}\sqrt{\langle x^2 \rangle \langle p_x^2 \rangle - \langle x p_x \rangle^2} \\
&= \frac{1}{mc}\sqrt{\langle x^2 \rangle \langle p_z^2 x'^2 \rangle - \langle x p_z x' \rangle^2} \\
&= \frac{1}{mc}\sqrt{\langle p_z^2 \rangle \langle x^2 \rangle \langle x'^2 \rangle - \langle p_z \rangle^2 \langle x x' \rangle^2} \\
&= \frac{1}{mc}\sqrt{\langle p_z^2 \rangle \langle x^2 \rangle \langle x'^2 \rangle - \langle p_z \rangle^2 \langle x^2 \rangle \langle x'^2 \rangle + \langle p_z \rangle^2 \langle x^2 \rangle \langle x'^2 \rangle - \langle p_z \rangle^2 \langle x x' \rangle^2} \\
&= \frac{\langle p_z \rangle}{mc}\sqrt{\hat{\sigma}_E^{\,2} \langle x^2 \rangle \langle x'^2 \rangle + \epsilon_{geo}^2}
\end{aligned}
\tag{3.1}
$$

where $\hat{\sigma}_E = \sqrt{\langle p_z^2 \rangle - \langle p_z \rangle^2}/\langle p_z \rangle$, and we also assume there is no correlation between p_z and x'.

Next we show the emittance growth in two specific scenarios: free space drifting and transport in a uniform focusing field.

3.2.1 Emittance Growth in Free Space Drifting

When a relativistic beam with finite energy spread propagates in free space, $x = x_0 + x_0' z$, $x' = x_0'$, then,

$$
\begin{aligned}
\epsilon_{geo} &= \sqrt{\langle x^2 \rangle \langle x'^2 \rangle - \langle x x' \rangle^2} \\
&= \sqrt{\langle x_0^2 \rangle \langle x_0'^2 \rangle - \langle x_0 x_0' \rangle^2} \\
&= \epsilon_{geo,0}
\end{aligned}
\tag{3.2}
$$

and,

$$
\begin{aligned}
\epsilon_n &= \frac{\langle p_z \rangle}{mc}\sqrt{\hat{\sigma}_E^2 \langle x_0^2 + 2 x_0 x'_0 z + x_0'^2 z^2 \rangle \langle x_0'^2 \rangle + \epsilon_{geo}^2} \\
&= \frac{\langle p_z \rangle}{mc}\sqrt{\hat{\sigma}_E^2 \epsilon_{geo}^2 \left(\beta_i - 2\alpha_i z + \gamma_i z^2 \right) \gamma + \epsilon_{geo}^2} \\
&= \frac{\langle p_z \rangle}{mc}\epsilon_{geo}\sqrt{\left(\hat{\sigma}_E^2 \left[(\gamma_i z - \alpha_i)^2 + 1 \right] + 1 \right)} \\
&= \epsilon_{n,0}\sqrt{\left(\hat{\sigma}_E^2 \left[(\gamma_i z - \alpha_i)^2 + 1 \right] + 1 \right)}
\end{aligned}
\tag{3.3}
$$

One can see the geometric emittance ϵ_{geo} remains invariant and the normalized emittance ϵ_n grows.

For an electron beam exiting from a plasma accelerator with $\beta_i = \sqrt{2\langle\gamma_b\rangle}k_p^{-1}$, $\alpha_i = 0$,

$$\epsilon_n(z) \approx \epsilon_{n,0} \frac{k_p z \hat{\sigma}_E}{\sqrt{2\langle\gamma_b\rangle}} \tag{3.4}$$

for $k_p z \gg \sqrt{2\langle\gamma_b\rangle}/\hat{\sigma}_E$. For an injector with a density around $10^{19}\,\mathrm{cm}^{-3}$, and the typical parameters of the beam ($\langle\gamma_b\rangle = 100, \hat{\sigma}_E = 0.01, \beta_i = 24\,\mu\mathrm{m}, \alpha_i = 0$), the normalized emittance grows to 4 folds of the initial value after 1 cm drift.

Physically, it is very simple to understand why this emittance growth occurs: for each energy group within the electron bunch, its transverse expansion rate is different due to the energy difference. Even though initially these groups have identical phase space profile, over certain distance they will occupy different area in the projected phase space, as a result the total normalized emittance grows as shown in Fig. 3.2a.

3.2.2 Emittance Growth in a Uniform Focusing Field

Next we estimate the emittance growth and the saturation value of the emittance in a mismatch uniform focusing field. When the electron beam propagates in quadrupoles or plasma wakes, if the initial phase space is matched with the focusing gradient, i.e., $\beta_i = \beta_F, \alpha_i = 0$, where $\beta_F = \sqrt{\langle\gamma_b\rangle\,mc/Ge}$ is the average beta function of the beam in the focusing element, the phase space of the beam remains invariant and the emittance stays constant; if not matched, the normalized and geometric emittance will grow to a saturation value as shown in Fig. 3.2b. The emittance evolution is determined by the detailed configurations of the quadrupoles or the field structure in the wake. For simple case where only linear focusing force with constant gradient is present and the longitudinal force is neglected, the saturation emittance can be

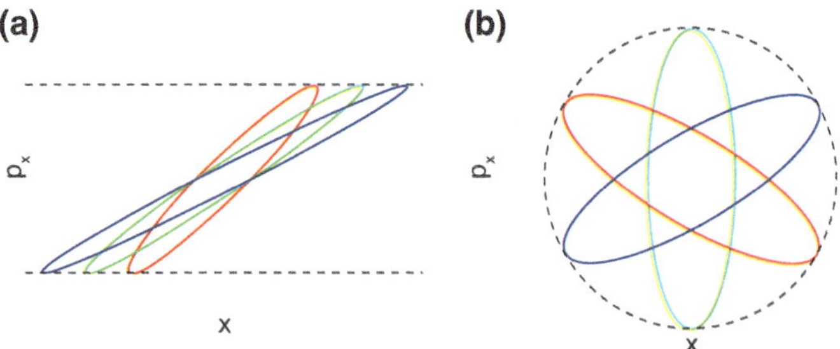

Fig. 3.2 The decoherence of the beam with finite energy spread. **a** In free space; **b** In quadrupoles. Note the color represents the energy

estimated by assuming the electrons fill in the phase space ellipse determined by the initial C-S parameters and the focusing gradient. The initial phase space ellipse of the electrons is

$$\gamma_i x^2 + 2\alpha_i x x' + \beta_i x'^2 = \epsilon_{geo,0} \tag{3.5}$$

which can be rewritten as

$$\gamma_i \beta_F \left(\frac{x}{\sqrt{\beta_F}}\right)^2 + 2\alpha_i \left(\frac{x}{\sqrt{\beta_F}}\right)\left(x'\sqrt{\beta_F}\right) + \frac{\beta_i}{\beta_F}\left(x'\sqrt{\beta_F}\right)^2 = \epsilon_{geo,0} \tag{3.6}$$

The long axis is

$$a = \sqrt{\frac{2\epsilon_{geo,0}}{(\gamma_i \beta_F + \beta_i/\beta_F) - \sqrt{(\gamma_i \beta_F - \beta_i/\beta_F)^2 + 4\alpha_i^2}}}$$

$$= \sqrt{\frac{2\epsilon_{geo,0}}{(\gamma_i \beta_F + \beta_i/\beta_F) - \sqrt{(\gamma_i \beta_F + \beta_i/\beta_F)^2 - 4}}} \tag{3.7}$$

Then, the saturated geometric emittance can be estimated as

$$\epsilon_{geo,sat} = a^2 \approx \frac{2\epsilon_{geo,0}}{(\gamma_i \beta_F + \beta_i/\beta_F) - \sqrt{(\gamma_i \beta_F + \beta_i/\beta_F)^2 - 4}} \tag{3.8}$$

and the saturated normalized emittance is

$$\epsilon_{n,sat} = \langle \gamma_b \rangle \epsilon_{geo,sat} \approx \frac{2\epsilon_{n,0}}{(\gamma_i \beta_F + \beta_i/\beta_F) - \sqrt{(\gamma_i \beta_F + \beta_i/\beta_F)^2 - 4}} \tag{3.9}$$

When $\gamma_i \beta_F + \beta_i/\beta_F \gg 4$, the saturated values can be estimated as

$$\epsilon_{n,sat} = \langle \gamma_b \rangle \epsilon_{geo,sat} \approx \epsilon_{n,0}\left(\gamma_i \beta_f + \frac{\beta_i}{\beta_f}\right) \tag{3.10}$$

The decoherence length which is defined as the emittance reaches its saturation value can be estimated by

$$L_{dc} \approx \frac{\pi}{\Delta_{\beta^{-1}}} \approx \frac{\pi \beta_f}{\hat{\sigma}_E} \tag{3.11}$$

For example, transporting an electron beam ($\langle \gamma_b \rangle = 2000$, $\hat{\sigma}_E = 0.01$, $\beta_i = 0.75\,\mathrm{mm}$, $\alpha_i = 0$) from a plasma (density $10^{17}\,\mathrm{cm}^{-3}$) to a quadrupole with $G = 1000\,\mathrm{T/m}$, the normalized emittance of the beam grows to $\sim 5\epsilon_{n0}$ after 1 m and

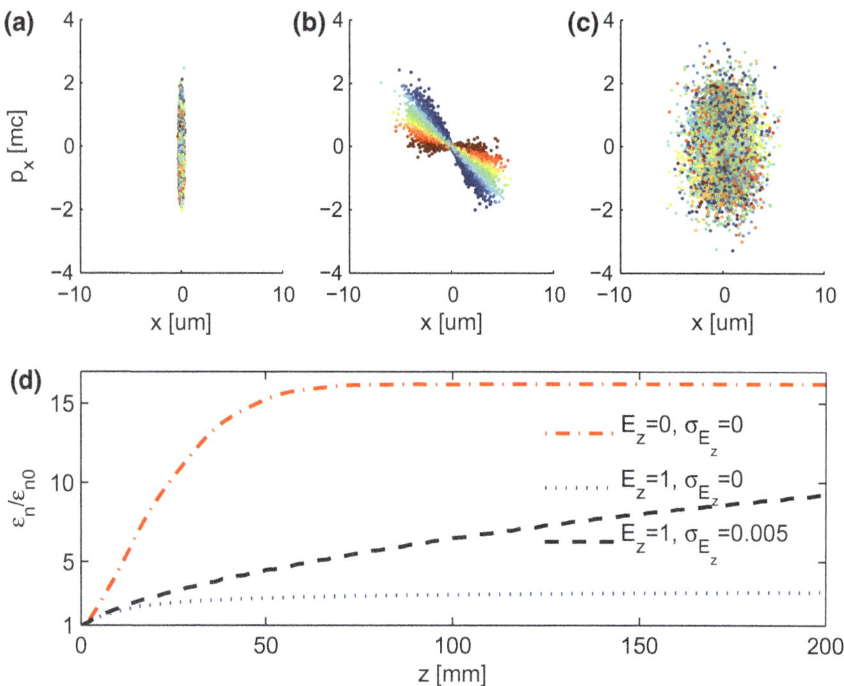

Fig. 3.3 The phase space and emittance evolution of an electron beam from a high density plasma injector ($10^{19}\,\mathrm{cm}^{-3}$) with $\langle \gamma_b \rangle = 200$, $\hat{\sigma}_E = 0.01$. It drifts $0.5\,\mathrm{mm}$ in vacuum and then propagates in a low density plasma accelerator ($10^{17}\,\mathrm{cm}^{-3}$). The $x - p_x$ phase space at **a** $z = 0\,\mathrm{mm}$, **b** $z = 12.5\,\mathrm{mm}$ and **c** $z = 200\,\mathrm{mm}$. The color represents the energy. **d** The emittance evolution under different acceleration gradients (normalized in $mc\omega_p/e$)

saturates at $\sim 30\epsilon_{n0}$ after $15\,\mathrm{m}$. We note the emittance growth due to finite energy spread is mitigated when the beam energy is much higher than the GeV level.

In Fig. 3.3 we also plot the emittance evolution for a two-stage plasma accelerator. An electron beam from a plasma injector ($10^{19}\,\mathrm{cm}^{-3}$) with $\langle \gamma_b \rangle = 200$, $\hat{\sigma}_E = 0.01$ drifts $0.5\,\mathrm{mm}$ in free space and then propagates into a low density plasma accelerator ($10^{17}\,\mathrm{cm}^{-3}$). The emittance grows to $16\epsilon_{n0}$ after $100\,\mathrm{mm}$ propagation in the low density plasma if no acceleration is taken into account, and this is in good agreement with the analytical calculations Eqs. 3.10 and 3.11. If there is acceleration, the emittance still saturates at the predicted value but with a different saturation length determined by the acceleration gradient E_z and the spread of the acceleration gradient σ_{E_z}. If the acceleration is uniform over the whole beam, the relative energy spread $\hat{\sigma}_E$ decreases as the energy is boosted, thus the betatron phase spread of the electron beam is frozen, and the emittance saturates at a lower value as shown by the dotted line in Fig. 3.3.

3.3 Theoretical Analysis of a Matching Plasma

As mentioned earlier, due to the drastic differences of the transverse focusing strength between PAs and the traditional accelerator components, the emittance of the beam grows quickly during transporting if finite energy spread is present. Here we propose to use longitudinally tailored plasma structure to properly guide the beam through stages while maintaining the beam quality. When a driver, intense laser pulse or relativistic electron beam, propagates in the plasma, the plasma electrons are blown out by the ponderomotive force of the laser or the electric field of the charged beam and an ion column is left. If the plasma density variation is adiabatic, i.e., $\left|(1/n_p)\mathrm{d}n_p/\mathrm{d}z\right| \ll k_p$, the quasi-static approximation is still valid, and the linear focusing force in the ion column is determined by the local plasma density. By carefully tailoring the density profile of the plasma, the electron beam phase space can be transformed into a proper profile to match with the next stage.

3.3.1 How to Design the Matching Plasma?

When an electron propagates in an ion column, its transverse motion equation is,

$$\frac{\mathrm{d}^2 x}{\mathrm{d}z^2} + K(z)x = 0 \tag{3.12}$$

where $K(z) = gn_p(z)e^2/\left(2\gamma_b mc^2\epsilon_0\right)$, and $n_p(z)$ is the local plasma density, γ_b is the relativistic factor of the electron, $g = 1$ indicated that there is no electron left in the ion column.

To get an analytical solution, we assume the density profile has a form as $n_p(z) = n_{p0}l^2/(z+l)^2$, where $0 \leq z \leq L$. Now Eq. 3.12 becomes an Eular's equation $\mathrm{d}^2 y/\mathrm{d}x^2 - Ax^\alpha y = 0$ with $\alpha = -2$, $A = -g\left(k_{p0}l\right)^2/2\gamma_b$. For an oscillation solution, the coefficient A of the Eular's equation should satisfy $4A + 1 < 0$, i.e., $k_{p0}l > \sqrt{\gamma_b/2g}$. When $k_{p0}l > \sqrt{\langle\gamma_b\rangle/2g}$, the analytical solutions of Eq. 3.12 are,

$$x = c_1\xi^{1/2}\cos(s\ln\xi) + c_2\xi^{1/2}\sin(s\ln\xi) \tag{3.13}$$

$$x' = c_1\xi^{-1/2}\left[\cos(s\ln\xi)/2 - s\sin(s\ln\xi)\right] + c_2\xi^{-1/2}\left[\sin(s\ln\xi)/2 + s\cos(s\ln\xi)\right] \tag{3.14}$$

where $\xi = z + l$, $s = \sqrt{g\left(k_{p0}l\right)^2/2\langle\gamma_b\rangle - 1/4}$, and c_1, c_2 are the coefficients determined by the initial conditions. We note that for this special density profile, the $K(z)$ is not necessarily to be adiabatic since the solutions here are exact.

Introducing the transfer matrix as $\begin{pmatrix} x \\ x' \end{pmatrix} = M(\xi|l)\begin{pmatrix} x_0 \\ x_0' \end{pmatrix}$, then the C-S parameters at $\xi = z + l$ can be expressed as [18]

$$\beta(\xi) = M_{11}^2 \beta_i - 2M_{11} M_{12} \alpha_i + M_{12}^2 \gamma_i \tag{3.15}$$

$$\alpha(\xi) = -M_{11} M_{21} \beta_i + (M_{11} M_{22} + M_{12} M_{21}) \alpha_i - M_{12} M_{22} \gamma_i \tag{3.16}$$

where,

$$M\left(\xi | l\right) = M\left(\xi | 0\right) M\left(l | 0\right)^{-1}$$

$$= \begin{pmatrix} \sqrt{\frac{\xi}{l}} \left[\cos\left(s\ln\frac{\xi}{l}\right) - \frac{1}{2s}\sin\left(s\ln\frac{\xi}{l}\right)\right] & \frac{\sqrt{\xi l}}{s}\sin\left(s\ln\frac{\xi}{l}\right) \\ -\frac{1+4s^2}{4s\sqrt{\xi l}}\sin\left(s\ln\frac{\xi}{l}\right) & \sqrt{\frac{l}{\xi}}\left[\cos\left(s\ln\frac{\xi}{l}\right) + \frac{1}{2s}\sin\left(s\ln\frac{\xi}{l}\right)\right] \end{pmatrix}$$

$$\tag{3.17}$$

where the determinant of $M\left(\xi | l\right)$ is unity. It is straightforward to obtain the betatron phase as $\phi(\xi) = s\ln(\xi/l) + \phi_0$, where ϕ_0 is the initial betatron phase.

For given initial C-S parameters of the beam and the targeted matching profile, one can numerically solve Eqs. 3.15 and 3.16 to obtain the characteristic parameters l and L of the matching plasma.

A numerical example is shown in Fig. 3.4. Due to the intrinsic periodical nature of the betatron oscillation, there are many solutions (l, L) for given $(\beta_i, \alpha_i, \beta_{goal}, \alpha_{goal})$, which are corresponding to $n_0 + N/2(N = 0, 1, 2, \ldots)$ betatron periods. The phase space trajectories of a test electron is shown in Fig. 3.4c and the evolution of the C-S parameters for $N = 0, 1, 2$ are shown in Fig. 3.4d.

3.3.2 The Effect of the Energy Spread

In this subsection we examine the performance of the matching plasma for an electron beam with finite energy spread. Considering a particle with energy $\gamma_b = \langle \gamma_b \rangle + \Delta_{\gamma_b}$, the betatron phase advance in the matching stage is

$$\Delta\phi(\gamma_b) = s(\gamma_b)\ln\frac{\xi}{l}$$

$$= \ln\frac{\xi}{l}\sqrt{\frac{g(k_{p0}l)^2}{2}\frac{1}{\gamma_b} - \frac{1}{4}}$$

$$\approx \ln\frac{\xi}{l}\sqrt{\frac{g(k_{p0}l)^2}{2}\frac{1}{\langle\gamma_b\rangle} - \frac{1}{4} - \frac{g(k_{p0}l)^2}{2}\frac{1}{\langle\gamma_b\rangle}\frac{\Delta_{\gamma_b}}{\langle\gamma_b\rangle}}$$

$$= \Delta\phi\left(\langle\gamma_b\rangle\right)\sqrt{1 - \frac{1}{1 - \frac{\langle\gamma_b\rangle}{2g(k_{p0}l)^2}}\frac{\Delta_{\gamma_b}}{\langle\gamma_b\rangle}} \tag{3.18}$$

One can see if $1 - \frac{\langle\gamma_b\rangle}{2g(k_{p0}l)^2}$ is not close to zero, the phase difference induced by the energy spread is not serious, i.e.,

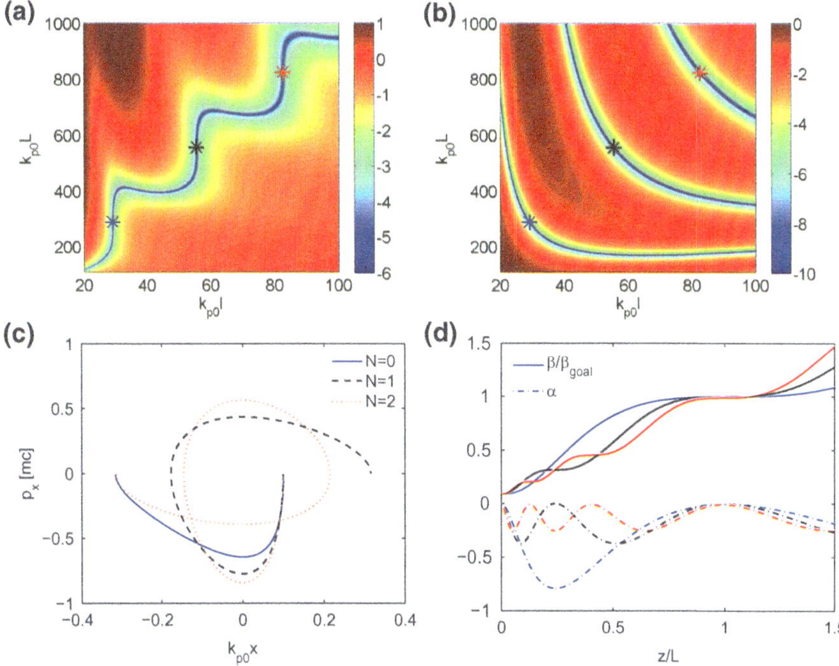

Fig. 3.4 The parameters are $\gamma_0 = 200$, $k_{p0}\beta_i = 20$, $\alpha_i = 0$, $k_{p0}\beta_{goal} = 200$, $\alpha_{goal} = 0$ and $g = 1$. **a** The logarithm of $\left|\beta(l, L) - \beta_{goal}\right|/\beta_{goal}$. **b** The logarithm of $\left|\alpha(l, L) - \alpha_{goal}\right|$. **c** The trajectories of a test particle with different (l, L). **d** The evolution of the C-S parameters. The solid lines are for the β-function and the dot-dashed lines are for the α-function. Different colors correspond to different solutions of (l, L) as shown in (**a**) and (**b**)

$$\Delta\phi(\gamma_b) - \Delta\phi\left(\langle\gamma_b\rangle\right) \approx \Delta\phi\left(\langle\gamma_b\rangle\right)\left(-\frac{1}{2}\frac{1}{1 - \frac{\langle\gamma_b\rangle}{2g(k_{p0}l)^2}}\right)\frac{\Delta_{\gamma_b}}{\langle\gamma_b\rangle} \tag{3.19}$$

where $1 - \frac{\langle\gamma_b\rangle}{2g(k_{p0}l)^2} \gg \frac{\Delta_{\gamma_b}}{\langle\gamma_b\rangle}$ is assumed.

In the matching stage, the betatron phase advance is typically on the order of π [depending on the choice of (l, L)], therefore the decoherence of the beam due to the energy spread is not serious. In Fig. 3.5a, we show the trajectory of a test particle with different energy; in Fig. 3.5b, the evolution of the C-S parameters and the emittance of a beam with finite energy spread is shown and compared with the beam without energy spread. We can see even with $\hat{\sigma}_E = 0.1$, the matching plasma still works well.

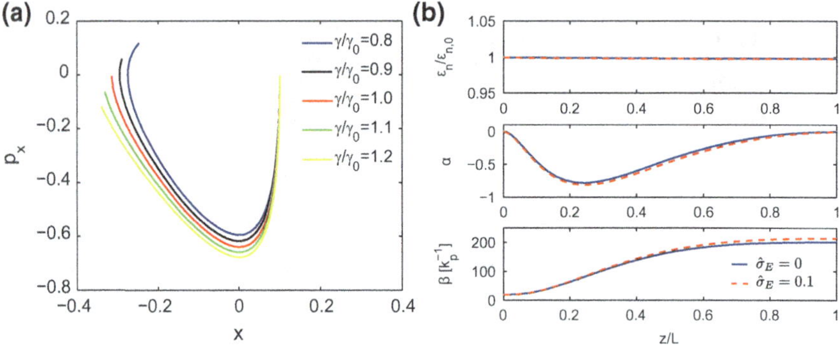

Fig. 3.5 **a** The trajectory of a test particle with different energy. **b** The evolution of the C-S parameters and the emittance of a beam with $\hat{\sigma}_E = 0.1$. The initial C-S parameters and the targeted matching profile are the same as the parameters in Fig. 3.4. The characteristic parameters l and L of the matching plasma with $N = 0$ is chosen

3.4 Verification by PIC Simulations

In this section, we give full 3D PIC simulation verification of three examples with longitudinally tailored plasma structures to match the electron beam between stages corresponding to Fig. 3.1a, b, c.

3.4.1 Matching Between Two-Stage LWFAs

First, we consider the matching between a high density plasma injector and a low density PA as shown in Fig. 3.1a. Assuming the plasma densities in the injector and accelerator are $n_{inj} = 10^{19}\,\mathrm{cm}^{-3}$ and $n_{acc} = 10^{17}\,\mathrm{cm}^{-3}$, respectively. An electron beam with $\langle \gamma_b \rangle = 200$, $\beta_i = 33.7\,\mu\mathrm{m}$, $\alpha_i = 0$ needs to be matched to $\beta_{goal} = 337\,\mu\mathrm{m}$, $\alpha_{goal} = 0$. By solving Eqs. 3.15 and 3.16 numerically, the parameters of the matching plasma are $l \approx 49\,\mu\mathrm{m}$, $L \approx 394\,\mu\mathrm{m}$. We note that the adiabatic condition is satisfied in this density profile.

This matching process is examined using the three-dimensional (3D) PIC code OSIRIS [19] in Cartesian coordinates using a moving window. We define the $\hat{z}$-axis to be the propagating direction of the drive laser. The simulation window has a dimension of $180k_0^{-1} \times 240k_0^{-1} \times 240k_0^{-1}$ with $900 \times 1200 \times 1200$ cells in the $\hat{x}$, $\hat{y}$ and $\hat{z}$ directions respectively, where k_0 is the wavenumber of the driver laser. The simulation results are shown in Fig. 3.6 and good agreements with the theoretical model are obtained.

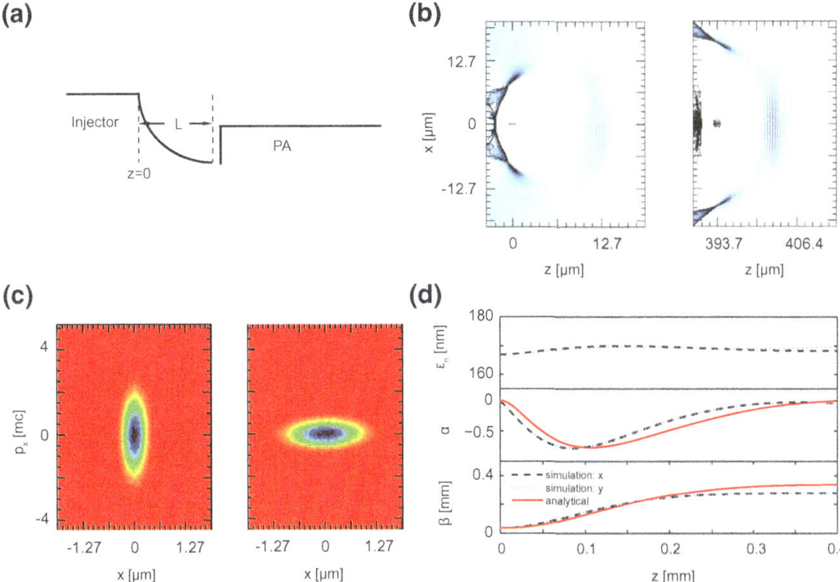

Fig. 3.6 **a** The configuration of the matching plasma between plasma based injector and PA. **b** The plasma and beam density distributions at $z = 0$ mm and $z = 0.39$ mm. **c** The $x - p_x$ phase space plots at $z = 0$ mm and $z = 0.39$ mm. **d** The emittance, β function and α function evolutions of the electron beam in the matching plasma. The parameters of the electron beam at $z = 0$ mm are: $\sigma_{x,y} = 0.17\,\mu$m, $\tau_{FWHM} = 5$ fs, $\langle \gamma_b \rangle = 200$, $\hat{\sigma}_E = 0.1$, $n_b = 10^{20}$ cm^{-3}. The 800 nm laser parameters are: $a_0 = 4$, $w_0 = 10\,\mu$m, $\tau_{FWHM} = 15$ fs. The laser is focused at $z = -0.04$ mm. The energy of the beam varies from $\langle \gamma_b \rangle = 200$, $\hat{\sigma}_E = 0.1$ at $z = 0$ mm to $\langle \gamma_b \rangle = 192$, $\hat{\sigma}_E = 0.105$ at $z = 0.39$ mm

3.4.2 Matching in External Injection

Next we consider the external injection case as shown in Fig. 3.1b. An electron beam generated from a RF-based accelerator is injected into a PA with $n_{acc} = 10^{17}$ cm^{-3}. If the C-S parameters is not matched with the plasma wake, the emittance would increase as the beam is accelerated in the plasma [16]. It is difficult to match the beam to the PA using the quadrupoles whose gradient is 3 orders of magnitude lower than that of a plasma wake with density 10^{17} cm^{-3}. Similar to the previous example, a longitudinally tailored plasma structure can be designed to match the beam from quadrupoles to the plasma wake. Due to the large simulation scale, 2D simulation results are used here. The simulation window has a dimension of $1600k_0^{-1} \times 3000k_0^{-1}$ with 8000×1500 cells in the $\hat{x}$ and $\hat{z}$ directions respectively. The initial beam parameters are set as $\langle \gamma_b \rangle = 50$, $\beta_i = 5$ mm, $\alpha_i = 0$ which can be realized in a traditional beam line. The C-S parameters in a 2D nonlinear wake are $\beta_{goal} = 0.12$ mm, $\alpha_{goal} = 0$. By solving Eqs. 3.15 and 3.16, the characteristic

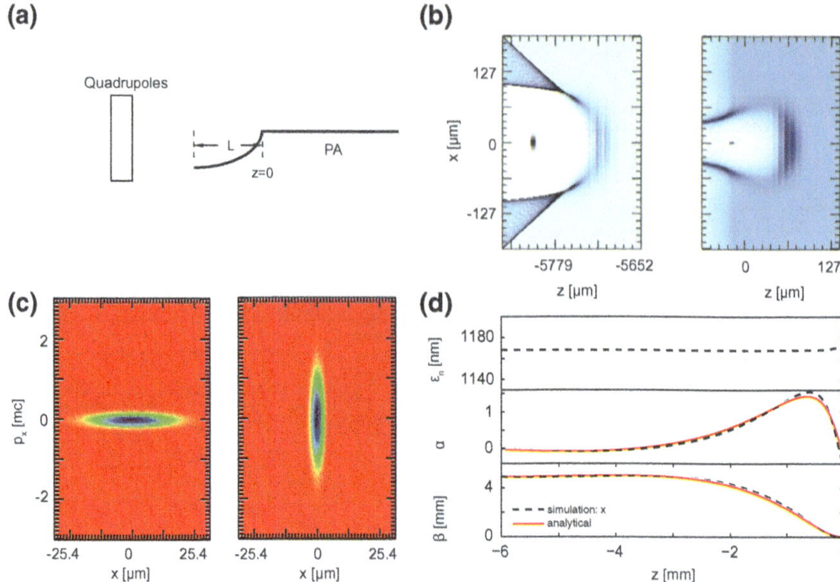

Fig. 3.7 **a** The configuration of the matching plasma between the RF-based accelerator and the PA. **b** The plasma and beam density distributions at $z = -5.8$ mm and $z = 0$ mm. **c** The $x - p_x$ phase space plots at $z = -5.8$ mm and $z = 0$ mm. **d** The emittance, β function and α function evolutions of the electron beam in the matching plasma. The parameters of the electron beam at $z = -5.8$ mm are: $\sigma_{x,y} = 10.9$ μm, $\tau_{FWHM} = 25$ fs, $\langle \gamma_b \rangle = 50$, $\hat{\sigma}_E = 0.02$, $n_b = 10^{16}$ cm^{-3}. The 800 nm laser parameters are: $a_0 = 3$, $w_0 = 2\sqrt{a_0}c/\omega_{p,acc} = 58$ μm, $\tau_{FWHM} = 100$ fs. The laser is focused at $z = 0$ mm. The energy of the beam varies from from $\langle \gamma_b \rangle = 50$, $\hat{\sigma}_E = 0.02$ at $z = -5.8$ mm to $\langle \gamma_b \rangle = 44.8$, $\hat{\sigma}_E = 0.0225$ at $z = 0$ mm

parameters of the plasma structure is $l \approx 0.116$ mm, $L \approx 5.8$ mm. The simulation results are shown in Fig. 3.7 and good agreements with the theoretical model are obtained.

3.4.3 Matching Between LWFAs and the Quadrupoles

Next we consider the beam coupling from the plasma accelerator to quadrupoles as shown in Fig. 3.1c. The 2D simulation window has a dimension of $1600k_0^{-1} \times 3000k_0^{-1}$ with 8000×3000 cells in the $\hat{x}$ and $\hat{z}$ directions respectively. The parameters of the electron beam generated from a 10^{17} cm^{-3} plasma accelerator are set as $\langle \gamma_b \rangle = 4000$, $\beta_i = 1.06$ mm, $\alpha_i = 0$, and the matching goal are $\beta_{goal} = 10.6$ mm, $\alpha_{goal} = 0$. By solving Eqs. 3.15 and 3.16, the characteristic parameters of the plasma structure is $l \approx 1.55$ mm, $L \approx 13.3$ mm. The simulation results are shown in Fig. 3.8 and good agreements with theoretical model are achieved.

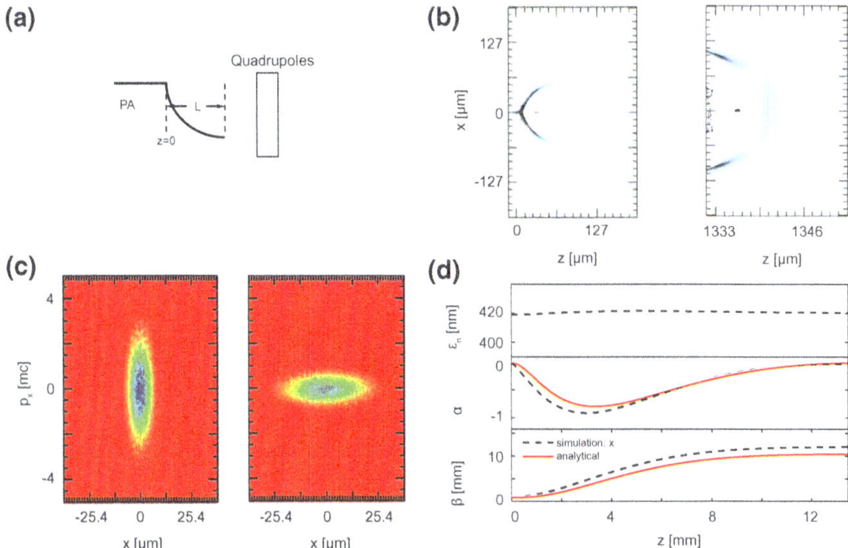

Fig. 3.8 **a** The configuration of the matching plasma between the PA and the quadrupoles. **b** The plasma and beam density distributions at $z = 0$ mm and $z = 13.3$ mm. **c** The $x - p_x$ phase space plots at $z = 0$ mm and $z = 13.3$ mm. **d** The emittance, β function and α function evolutions of the electron beam in the matching plasma. The parameters of the electron beam at $z = 0$ mm are: $\sigma_{x,y} = 0.34$ μm, $\tau_{FWHM} = 25$ fs, $\langle \gamma_b \rangle = 4000$, $\hat{\sigma}_E = 0.05$, $n_b = 10^{18}$ cm^{-3}. The 800 nm laser parameters are: $a_0 = 3$, $w_0 = 2\sqrt{a_0}c/\omega_{p,acc} = 58$ μm, $\tau_{FWHM} = 100$ fs. The laser is focused at $z = 0$ mm. The energy of the beam varies from from $\langle \gamma_b \rangle = 4000$, $\hat{\sigma}_E = 0.05$ at $z = 0$ mm to $\langle \gamma_b \rangle = 3966.6$, $\hat{\sigma}_E = 0.0506$ at $z = 13.3$ mm

We note that for the coupling between stages as shown in Fig. 3.1d, one can combine the plasma structures in Figs. 3.7 and 3.8 to transport the electron beam without degradation of the beam quality.

3.5 Summary

For the further development of plasma based accelerators, phase space matching between plasma acceleration stages and between plasma stages and traditional accelerator components becomes a very critical issue for high quality high energy acceleration and its applications in light sources and colliders. Without proper matching, catastrophic emittance growth in the presence of finite energy spread may occur when the beam propagating through different stages and components due to the drastic differences of transverse focusing strength. In this chapter we propose to use longitudinally tailored plasma structures as phase space matching components to properly guide the beam through stages. Theoretical analysis and particle-in-cell simulations are utilized to show clearly how these structures may work in four different scenarios. Very good agreements between theory and simulations are obtained.

References

1. Leemans W, Esarey E (2009) Laser-driven plasma-wave electron accelerators. Phys Today 62:44
2. Adli E, Delahaye JP, Gessner SJ et al (2013) A beam driven Plasma-Wakefield linear collider: from Higgs factory to multi-TeV. arXiv:1308.1145
3. Schlenvoigt HP, Haupt K, Debus A et al (2007) A compact synchrotron radiation source driven by a laser-plasma wakefield accelerator. Nat Phys 4(2):130–133
4. Fuchs M, Weingartner R, Popp A et al (2009) Laser-driven soft-X-ray undulator source. Nat Phys 5(11):826–829
5. Kneip S, McGuffey C, Martins JL et al (2010) Bright spatially coherent synchrotron X-rays from a table-top source. Nat Phys 6(12):980–983
6. Cipiccia S, Islam MR, Ersfeld B et al (2011) Gamma-rays from harmonically resonant betatron oscillations in a plasma wake. Nat Phys 7(11):867–871
7. Gonsalves A, Nakamura K, Lin C et al (2011) Tunable laser plasma accelerator based on longitudinal density tailoring. Nat Phys 7(11):862–866
8. Liu JS, Xia CQ, Wang WT et al (2011) All-optical cascaded laser wakefield accelerator using ionization-induced injection. Phys Rev Lett 107:035001
9. Pollock BB et al (2011) Demonstration of a narrow energy spread, $\sim$0.5 GeV electron beam from a two-stage laser wakefield accelerator. Phys Rev Lett 107:045001
10. Kim HT, Pae KH, Cha HJ et al (2013) Enhancement of electron energy to the multi-GeV regime by a dual-stage laser-wakefield accelerator pumped by petawatt laser pulses. Phys Rev Lett 111:165002
11. Stragier XFD, Luiten OJ, Geer SB et al (2011) Experimental validation of a radio frequency photogun as external electron injector for a laser wakefield accelerator. J Appl Phys 110(2):024910
12. Rossi AR, Bacci A, Belleveglia M et al (2014) The external-injection experiment at the SPARC_LAB facility. Nucl Instrum Methods Phys Res, Sect A 740(0):60–66. Proceedings of the first European Advanced Accelerator Concepts Workshop 2013
13. Eichner T, Grüner F, Becker S et al (2007) Miniature magnetic devices for laser-based, table-top free-electron lasers. Phys Rev ST Accel Beams 10:082401
14. Harrison J, Joshi A, Lake J et al (2012) Surface-micromachined magnetic undulator with period length between $10\,\mu m$ and 1 mm for advanced light sources. Phys Rev ST Accel Beams 15:070703
15. Antici P, Bacci A, Benedetti C et al (2012) Laser-driven electron beamlines generated by coupling laser-plasma sources with conventional transport systems. J Appl Phys 112(4):044902
16. Mehrling T, Grebenyuk J, Tsung FS et al (2012) Transverse emittance growth in staged laser-wakefield acceleration. Phys Rev ST Accel Beams 15:111303
17. Migliorati M, Bacci A, Benedetti C et al (2013) Intrinsic normalized emittance growth in laser-driven electron accelerators. Phys Rev ST Accel Beams 16:011302
18. Lee SY (1999) Accelerator physics. World Scientific, Singapore
19. Fonseca RA, Silva LO, Tsung FS et al (2002) OSIRIS: a three-dimensional, fully relativistic particle in cell code for modeling plasma based accelerators. Lecture notes in computer science, vol 2331, pp 342–351

Chapter 4
X-FELs Driven by Plasma Based Accelerators

4.1 Introduction

Since the first soft X-ray FEL FLASH lasing at Deutsches Elektronen-Synchrotron (DESY) in 2007 [1] and the first hard X-ray FEL LCLS (Linac Coherent Light Source) lasing at SLAC in 2010 [2], these machines have made great contributions to many branches of research in physics, chemistry, biology, materials and environmental science. Due to their short wavelength, high coherence and high power, these laser-like X-ray sources are revolutionary tools for many scientific disciplines. To drive a X-FEL, high quality electron beams with multi-GeV energy are needed, and this needs a huge investment on large-scale accelerators and undulators. Up to now, only a few X-FEL machines have been build around the world [1–5].

With the rapid progress of plasma based accelerators, people start to think about a compact X-FEL driven by plasma accelerators seriously now. Due to its unique process of electron beam generation, acceleration and transportation, the parameters space of the electron beams from plasma accelerators can be quite different with the beams from RF-based accelerators. Beams from plasma accelerators typically have ultrashort pulse length, very high peak current, relatively low emittance and high beam density, at the same time also have relatively large energy spread. Therefore, how can a compact X-FEL based on plasma acceleration work is still an open question and many challenges remains to be addressed. In this chapter, we will present some preliminary study to address these issues.

4.1.1 The Basic Principles of FELs

The lasing principle of FELs [6] is totally different compared with conventional lasers which utilize the population inversion between energy levels in atoms or molecules [7]. In FELs, a relativistic electron beam are sent into an undulator. The electrons oscillate transversely in the undulator and radiate. The spontaneous radiation can

© Springer Nature Singapore Pte Ltd. 2020

X. Xu, *Phase Space Dynamics in Plasma Based Wakefield Acceleration*,
Springer Theses, https://doi.org/10.1007/978-981-15-2381-6_4

bunch the electrons at the resonant wavelength and the bunched electrons can emit more photons coherently, and this leads to an instability between the electron beam and the radiation. The radiated power grows exponentially and then saturates [8]. This kind of FELs starting from the spontaneous radiation is called self amplified spontaneous emission (SASE) [9, 10]. The most X-FELs work in this mode.

The resonant wavelength in FELs is

$$\lambda_r = \frac{\lambda_u}{2\langle\gamma_b\rangle^2}(1 + a_u^2) \tag{4.1}$$

where λ_u is the wavelength of the undulator, $\langle\gamma_b\rangle$ is the relativistic factor of the electron beam and a_u is the normalized strength of the undulator with $a_u = eB_p\lambda_u/2\sqrt{2}\pi mc$ for a planar undulator and $a_u = eB_p\lambda_u/2\pi mc$ for a helical undulator, where B_p is the on-axis peak magnetic field. The typical parameters of static-magnetic undulators are $\lambda_u \sim$ cm and $a_u \sim 1$. One can see the beam energy needs to be several GeV to radiate X-rays.

In the 1D model where the electron beam has uniform transverse distribution with zero emittance and energy spread, the gain length which is defined as the length where the radiated power grows e fold can be expressed as

$$L_{g,1D} = \frac{\lambda_u}{4\pi\sqrt{3}\rho} \tag{4.2}$$

In the SASE FELs, saturation occurs after about 20 gain lengths, and the saturation power is

$$P_{sat,1D} \approx \rho P_{beam} \tag{4.3}$$

where $P_{beam} = \langle\gamma_b\rangle mc^2 I_p$ is the electron beam power, and ρ is the Pierce parameter which is defined as

$$\rho = \frac{1}{\gamma_b}\left[\left(\frac{a_u f_c \lambda_u}{8\pi\sigma_b}\right)^2 \frac{I_p}{I_A}\right]^{1/3} \tag{4.4}$$

where σ_b is the beam radius, I_p is the peak current and $I_A = 17$ kA, f_c is the coupling factor of the electrons to the radiation field due to the longitudinal wiggle motion with $f_c = J_0\left(a_u^2/2(1 + a_u^2)\right) - J_1\left(a_u^2/2(1 + a_u^2)\right)$ for a planar undulator and $f_c = 1$ for a helical undulator, where $J_0(x)$, $J_1(x)$ are Bessel functions. The Pierce parameter ρ is a critical parameter in the FEL, it is also called the FEL parameter.

The longitudinal bunching of the electrons is the key process in the FELs. The spread of forward velocities can destroy the bunching, therefore the FELs have stringent requirements of the beam emittance and energy spread. The geometric emittance of the beam must be smaller than the radiated wavelength, i.e.,

$$\epsilon_n < \langle\gamma_b\rangle\lambda_r \tag{4.5}$$

where ϵ_n is the normalized emittance of the beam. And the energy spread must be small than the Pierce parameter ρ, i.e.,

$$\frac{\sigma_\gamma}{\langle \gamma_b \rangle} < \rho \tag{4.6}$$

In order to quantify the effect of emittance, energy spread and other high-dimensional effects, Ming Xie gave a scaling law by fitting numerical solutions of the coupled Maxwell–Vlasov equations describing FEL interactions [11]. In this scaling law, the 3D gain length is

$$L_g = L_{g,1D}(1 + \eta) \tag{4.7}$$

where

$$\eta = a_1 \eta_d^{a_2} + a_3 \eta_\epsilon^{a_4} + a_5 \eta_\gamma^{a_6} + a_7 \eta_\epsilon^{a_8} \eta_\gamma^{a_9} + a_{10} \eta_d^{a_{11}} \eta_\gamma^{a_{12}} + a_{13} \eta_d^{a_{14}} \eta_\epsilon^{a_{15}} + a_{16} \eta_d^{a_{17}} \eta_\epsilon^{a_{18}} \eta_\gamma^{a_{19}} \tag{4.8}$$

and

$$\eta_d = \frac{L_{g,1D}}{L_r}, \quad \eta_\epsilon = \frac{L_{1D}}{\langle \beta \rangle} \frac{4\pi \epsilon_n}{\lambda_r \langle \gamma_b \rangle}, \quad \eta_\gamma = 4\pi \frac{L_{g,1D}}{\lambda_u} \frac{\sigma_\gamma}{\langle \gamma_b \rangle} \tag{4.9}$$

$L_r = 4\pi \sigma_b^2 / \lambda_r$ is the Rayleigh length of the radiation and $\langle \beta \rangle$ is the average β function of the beam. The fitting parameters are

$$
\begin{aligned}
a_1 &= 0.45, \quad a_2 = 0.57, \quad a_3 = 0.55, \quad a_4 = 1.6, \quad a_5 = 3, \quad a_6 = 2, \quad a_7 = 0.35, \\
a_8 &= 2.9, \quad a_9 = 2.4, \quad a_{10} = 51, \quad a_{11} = 0.95, \quad a_{12} = 3, \quad a_{13} = 5.4, \\
a_{14} &= 0.7, \quad a_{15} = 9, \quad a_{16} = 1140, \quad a_{17} = 2.2, \quad a_{18} = 2.9, \quad a_{19} = 3.2
\end{aligned} \tag{4.10}
$$

The saturation power in this scaling law is

$$P_{sat} \approx \frac{1.6}{(1 + \eta)^2} \rho P_{beam} \tag{4.11}$$

The η_d represents the gain reduction due to 3D diffraction, η_ϵ and η_γ represent the gain reduction due to the longitudinal velocity spread of the electrons caused by the emittance and energy spread, respectively. We can use this scaling law to guide the optimization of the electron beam generation in plasma based accelerators.

4.1.2 The Challenges and Opportunities of X-FELs Driven by Plasma Based Accelerators

X-FELs have stringent requirements on the beam quality. To make this clear, we list the main parameters of LCLS in Table. 4.1. Comparing with the beam parameters

Table 4.1 The main parameters of LCLS

Parameter	Soft X-ray FEL	Hard X-ray FEL	Unit
Electrons			
E_b	3.5–6.7	13.6	GeV
ϵ_n	0.4	0.4	μm
I_p	0.5–3.5	2.5–3.5	kA
RMS slice energy spread $\sigma_\gamma / \langle \gamma_b \rangle$	0.07	0.04	%
Undulator			
λ_u	0.03	0.03	m
a_u	2.62	2.62	
X-rays			
L_g	$\sim$1.5	3.5	m
λ_r	0.6–2.2	0.15	nm
P_{sat}	3–35	15–40	GW
ρ	0.0017		

from plasma based accelerators listed in Table. 1.1, one can see two major differences: on one hand the beam from plasma accelerators typically has high current and low emittance, therefore its brightness can be as good or even much better than that of the LCLS beam; on the other hand the beam from plasma accelerators typically has large energy spread on the percent level which is much larger than that of the LCLS beam. Therefore to reduce the energy spread in plasma accelerators is important to drive a compact X-FEL.

Recently, two methods are proposed to reduce the requirement on the energy spread for FELs. Huang et al. proposed to use transverse gradient undulator (TGU) to compensate the effects of the beam energy spread. Such a transverse gradient undulator together with a properly dispersed beam can greatly reduce the effects of the beam energy spread on FEL performance [12, 13]. Another group proposed to decompress the beam to reduce the slice energy spread [14, 15]. However both methods reduce the requirement on the beam energy spread with a cost of beam density dilution. In this chapter, we adopt a two-stage LWFA scheme to reduce the energy spread [16–18]. A low energy electron beam is generated first in a high density injector and then be boosted in energy using a low density accelerator. Due to the long plasma wavelength in the second stage, the acceleration phase occupied by the injected beam is much narrowed, leading to a reduced energy spread.

4.2   X-FEL Driven by a Two-Stage LWFA

We propose to use a two-stage LWFA to generate high brightness low energy spread electron beam to drive an X-FEL. In the injection stage, an ultra-short ($\sim$10 fs), high current ($\sim$10 kA), low emittance ($\sim$50 nm) electron beam with low absolute energy spread ($\sigma_\gamma \sim 5$) is generated in a laser driven density downramp injection scheme ($\sim 10^{19}\,\mathrm{cm}^{-3}$). Then this high quality low energy ($\langle \gamma_b \rangle \sim 100$) electron beam is sent to a low density laser driven accelerator ($\sim 10^{17}\,\mathrm{cm}^{-3}$) and boosted to several GeV. After the acceleration, the relative energy spread is reduced to ~ 0.004. Then, the high quality beam is sent into a magnetic undulator and radiates. In this compact X-FEL driven by a two-stage LWFA, the electron beam needs to propagate between the plasma wakes with different focusing gradients and between the plasma wake and the undulator, therefore the transverse phase space matching while conserving the beam emittance is a critical issue. We use the longitudinal tailored plasma structure proposed in Sect. 3 to match the beam between stages.

4.2.1 Simulation of the Injector Stage

To generate a high brightness electron beam, a laser driven density downramp injection is used [18, 19]. An 800 nm laser pulse with $a_0 = 4$, $w_0 = 6.7\,\mu\mathrm{m}$ and $\tau_{FWHM} = 15\,\mathrm{fs}$ propagates into a plasma and excites a nonlinear blowout wake. At the beginning of the plasma, the density increases linearly from 0 to $1.5 \times 10^{19}\,\mathrm{cm}^{-3}$ during a $38\,\mu\mathrm{m}$ up-ramp as shown in Fig. 4.1a. After a $25.4\,\mu\mathrm{m}$ platform the plasma density decreases linearly from $1.5 \times 10^{19}\,\mathrm{cm}^{-3}$ to $10^{19}\,\mathrm{cm}^{-3}$ in $28\,\mu\mathrm{m}$ length. Note that the laser is focused at the end of the up-ramp. The simulation window has a dimension of $40.6 \times 40.6 \times 30.5\,\mu\mathrm{m}$ with $1600 \times 1600 \times 1200$ cells in the $\hat{x}$, $\hat{y}$ and $\hat{z}$ directions, respectively. This corresponds to cell sizes of $0.2k_0^{-1}$ in the three directions. The parameters of the generated beam at $z \approx 240\,\mu\mathrm{m}$ are shown in Fig. 4.1b, c. By tracking the injected electrons, we confirm that the electrons are trapped in the density downramp region. After the injector, the electron beam is sent into a low density plasma accelerator. The transverse phase space mismatch would lead to a catastrophic emittance growth in the presence of finite energy spread. Therefore we add a longitudinal tailored matching plasma at the end of the injector to expand the beam radius and reduce the angle divergence. By solving Eqs. 3.15 and 3.16 numerically, the characteristic parameters of the matching plasma are $l \approx 37\,\mu\mathrm{m}$, $L \approx 320\,\mu\mathrm{m}$. The density profile of the plasma in the injector is shown as Fig. 4.1a. We can see the C-S parameters of the beam are matched with the accelerator stage after the matching plasma, i.e., the β-functions are expanded by 10 folds and the α-functions are still close to zero.

After the matching plasma, the brightness of the beam stays the same as shown in Fig. 4.1d. However the slice energy spread grows. And this may due to the longitudinal space oscillation within the beam ($n_b \sim 10^{22}\,\mathrm{cm}^{-3}$ before the matching). Further studies are needed to explain this growth of the slice energy spread. From Fig. 4.1b, one can see the latter half of beam with high brightness is suitable for an X-FEL.

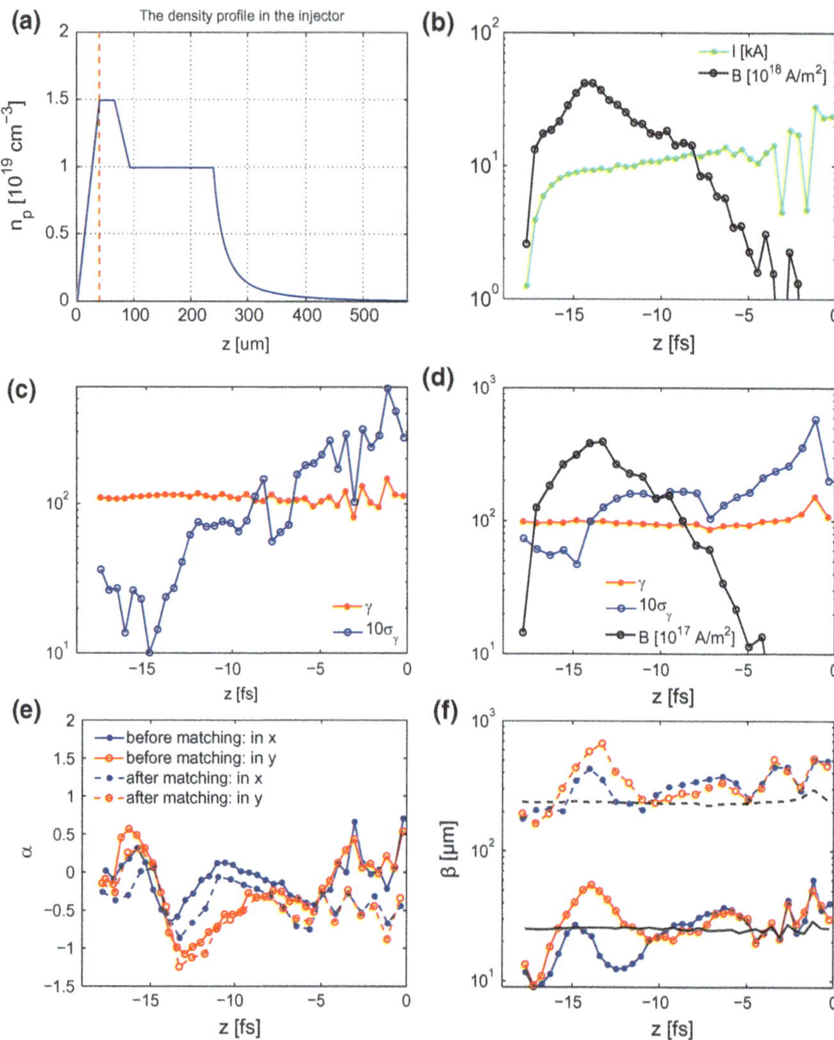

Fig. 4.1 a The plasma density profile in the injector. Note that the laser is focused at the end of the upramp, i,e., the dashed red line position. (**b**) and (**c**) Shows the beam parameters before the matching plasma. **d** Shows the beam parameter after the matching plasma. **e** and **f** Show the C-S parameters of the beam before and after the matching plasma. The black solid line shows the reference β-function $\beta = \sqrt{2\gamma_b}c/\omega_{p,inj}$, and the black dashed line is $\beta = \sqrt{2\gamma_b}c/\omega_{p,acc}$

4.2.2   Simulation of the Accelerator Stage

In this subsection we will show the feasibility to reduce the slice energy spread using a low density plasma wakefield accelerator with full-scale PIC simulations. However the simulation of high energy and high quality acceleration is very computationally intensive. To boost the accelerated beam to several GeV, the acceleration distance should be on the order of 0.1 m. And the grid size in the longitudinal direction needs to resolve the laser wavelength, therefore millions time steps are required to finish the simulation. The transverse radius of the beam is on the order of $0.1\,\mu$m due to the ultra low emittance ($\epsilon_n \sim 50$ nm), at the same time the plasma skin depth in the accelerator is $\sim 10\,\mu$m, therefore several hundreds of grids are needs in the transverse directions. In order to reduce the simulation scale and show the feasibility of this two-stage idea, we present the simulations of an accelerated beam with $\epsilon_n = 500$ nm in the following content.

The simulations are done utilizing a hybrid algorithm [20] which uses a PIC description in $r - z$ and a gridless description in the azimuthal direction. This algorithm can much reduce the simulation time. At the same time boosted frame technique [21, 22] is utilized to further reduce the required cpu-hours.

A beam with similar parameters as the beam generated in the injector is initialized. The normalized emittance of this accelerated beam is chosen as $\epsilon_n = 500$ nm with a corresponding width $\sigma_r = \sqrt{\beta_{acc}\epsilon_n/\langle\gamma_b\rangle} \approx 1.1\,\mu$m. The beam has a rectangle longitudinal profile with $\tau = 18$ fs. The initial mean energy and the energy spread of the accelerated beam are set as $\langle\gamma_b\rangle = 100, \sigma_\gamma = 10$. An 800 nm laser pulse with $w_0 = 56\,\mu$m, $\tau = 79$ fs propagates into the plasma ($n_{p,acc} = 10^{17}\,\mathrm{cm}^{-3}$). To avoid self-injection in the acceleration stage, the normalized vector potential of the driver laser is chosen as $a_0 = 3$ [23]. A plasma channel is set to externally guide the laser pulse. The channel profile is

$$n_p(r) = \begin{cases} n_{p,acc}\left(1 + 5r^2/r_0^2\right), & \text{when } r \leq r_0 \\ 6n_{p,acc}, & \text{when } r > r_0 \end{cases} \tag{4.12}$$

where $r_0 = 216\,\mu$m is the transverse size of the plasma channel.

After 0.1 m acceleration, the energy of the accelerated beam is boosted to 5 GeV. The evolution of the mean energy and the energy spread along the acceleration distance and the beam parameters at the exit of the accelerator are shown in Fig. 4.2. One can see as the energy is boosted, the slice energy energy spread also grows. Finally, the relative slice energy spread is 0.0037. The growth of the slice energy spread is probably due to the longitudinal space charge interaction between the high density electrons. One can see the final slice energy spread is mainly determined by this growth. At the same time, the normalized emittance stays constant as shown in Fig. 4.2e.

The β function of the accelerated beam at the exit of the accelerator is $\beta_{acc,end} = \sqrt{2\langle\gamma_b\rangle}c/\omega_{p,acc} \approx 2.4 \times 10^{-3}$ m. At the same time, the β function of the FODO structure along the LCLS undulator is on the order of ten meter. Therefore the match-

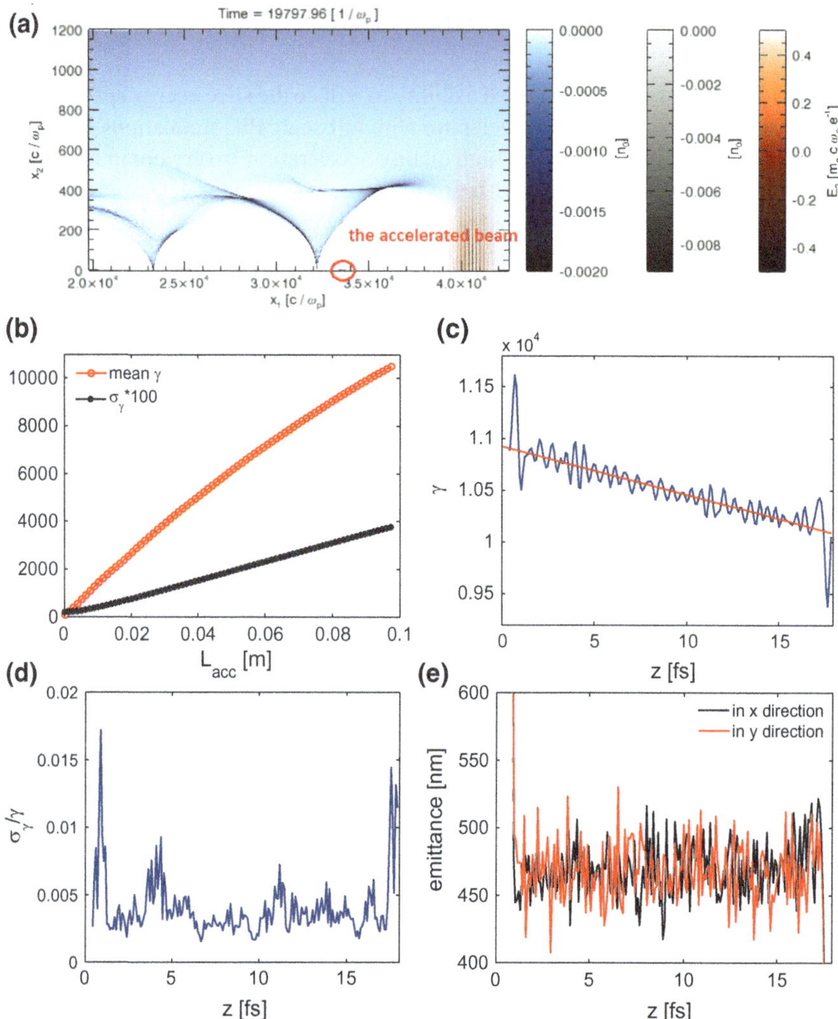

Fig. 4.2 a The concept plot . **b** The evolution of the mean energy and the energy spread of the accelerated beam. The energy distribution (**c**), the relative energy spread (**d**) and the emittance (**e**) of the accelerated beam at the end of the acceleration stage $z = 9.7$ cm

ing of the transverse space for the accelerated beam is required between the plasma accelerator and the undulator. The longitudinal tailored plasma structure can also be used to enlarge the β function of the accelerated beam in this situation. In the future, a start-to-end simulation which contains the required matching sections will be presented.

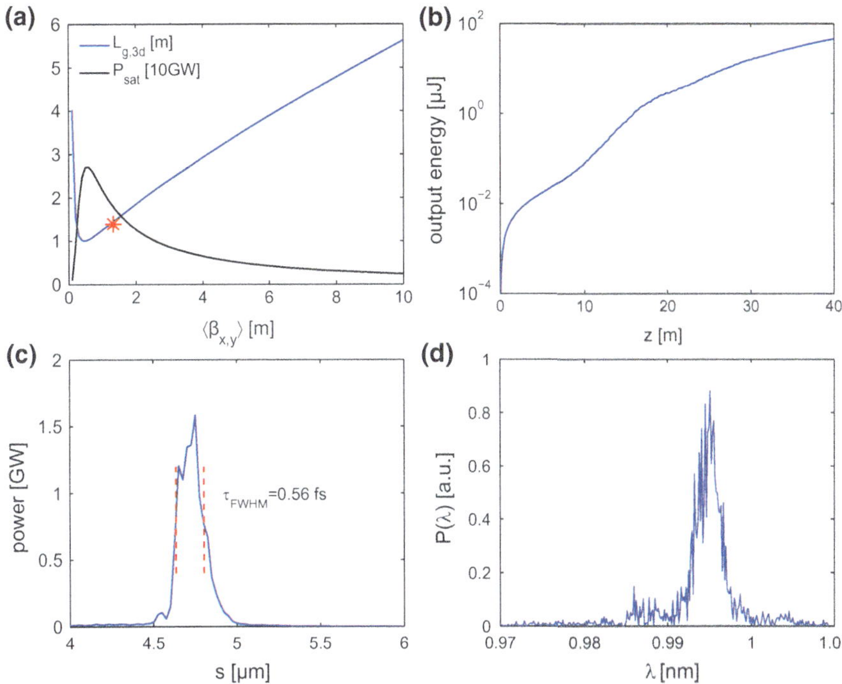

Fig. 4.3 **a** The dependence of the 3d gain length $L_{g,3d}$ and the saturation power on the average β function. The red star is the working point in the following simulations. **b** The output energy along the undulator. Note that this data is averaged over 11 simulations. **c** A typical X-ray pulse profile at $z = 18\,\text{m}$. Note that $s = 0\,\mu\text{m}$ represents the tail of the electron beam. **d** A typical spectra of the output X-ray at $z = 18\,\text{m}$

4.2.3 Simulation of the Undulator Stage

The data generated from OSIRIS is used by Genesis 1.3 [24] to simulate the radiation process in the undulator. The longitudinal phase space distribution of the beam from OSIRIS is retained in the undulator simulation. An ideal matching of the transverse phase space is assumed. In the simulation we use the parameters of the LCLS undulator, i.e., $\lambda_u = 0.03\,\text{m}$, $a_u = 2.62$. The dependence of the 3d gain length $L_{g,3d}$ and the saturation power on the average β-function is shown in Fig. 4.3a. One can see when $\langle\beta\rangle \approx 0.5\,\text{m}$ the shortest 3d gain length is obtained and the output power achieves its maximum. Therefore a FODO structure along the undulator which can achieve a meter-scale average β function is set in the simulation. The parameters of the external FODO structure are $G = 700\,\text{T/m}$, $l_F = l_D = 0.06\,\text{m}$ and $l_O = 0.6\,\text{m}$, which corresponds to $\beta_F = 2.25\,\text{m}$, $\beta_D = 0.37\,\text{m}$ for the beam generated from the plasma accelerator. The output energy of the X-ray along the undulator is shown in Fig. 4.3b. The output energy saturates around $z = 20$ m. The saturation power is much lower than the prediction from Ming's scaling law. The reason is due to the oscillation of the average beam energy as shown in Fig. 4.2c. The slippage length in this case is estimated as $L_s \approx 0.6\,\mu$m which is comparable with the beam length. Therefore the oscillation of the average beam energy is equivalent to a large slice energy spread, which decrease the saturation power. The X-ray pulse profile and the spectrum at $z = 18$ m are shown in Fig. 4.3c, d. One can see a giga-watt sub-fs X-ray pulse is generated in the undulator.

We note that the linear energy chirp in the longitudinal phase space of the accelerated beam (as shown by the red line in Fig. 4.2b) is removed in the above Genesis simulations. In plasma based acceleration, the linear energy chirp may be removed by fine tuning the longitudinal location and the current profile of the accelerated beam [25, 26].

4.3 Conclusions

In this chapter, we propose to drive a compact X-FEL using a two-stage LWFA. A high brightness electron beam is generated in a high density plasma using density ramp injection. Then this beam is sent into a low density accelerator and boosted to several GeV. Matching plasma structure is utilizing to match the injector and the accelerator. Finally, a high brightness and low slice energy spread electron beam is generated. Genesis 1.3 is used to simulate the X-FEL driven by such an electron beam.

References

1. Ackermann W, Asova G, Ayvazyan V et al (2007) Operation of a free-electron laser from the extreme ultraviolet to the water window. Nat Photon 1(6):336–342
2. Emma P, Akre R, Arthur J et al (2010) First lasing and operation of an ångstrom-wavelength free-electron laser. Nat Photon 4(9):641–647
3. Ishikawa T et al (2012) A compact X-ray free-electron laser emitting in the sub-angstrom region. Nat Photon 6(8):540–544
4. http://www.xfel.eu/en/
5. http://www.psi.ch/swissfel/
6. Madey J (1971) Stimulated emission of bremsstrahlung in a periodic magnetic field. J Appl Phys 42(5):1906–1931
7. Siegman AE (1986) Lasers. University Science Books, Mill Valley
8. Huang Z, Kim KJ (2007) Review of x-ray free-electron laser theory. Phys Rev ST Accel Beams 10:034801
9. Kondratenko AM, Saldin EL (1980) Generation of coherent radiation by a relativistic electron beam in an ondulator. Part Accel 10:206–16
10. Bonifacio R, Pellegrini C, Narducci L (1984) Collective instabilities and high-gain regime in a free-electron laser. Opt Commun 50(6):373–378
11. Xie M (1995) Design optimization for an X-ray free electron laser driven by SLAC linac. In: Proceedings of particle accelerator conference, Proceedings of the 1995, vol 1. IEEE, pp 183–185

12. Huang Z, Ding Y, Schroeder CB (2012) Compact X-ray free-electron laser from a laser-plasma accelerator using a transverse-gradient undulator. Phys Rev Lett 109:204801
13. Baxevanis P, Ding Y, Huang Z et al (2014) 3D theory of a high-gain free-electron laser based on a transverse gradient undulator. Phys Rev ST Accel Beams 17:020701
14. Maier AR, Meseck A, Reiche S et al (2012) Demonstration scheme for a laser-plasma-driven free-electron laser. Phys Rev X 2:031019
15. Seggebrock T, Maier AR, Dornmair I et al (2013) Bunch decompression for laser-plasma driven free-electron laser demonstration schemes. Phys Rev ST Accel Beams 16:070703
16. Liu JS, Xia CQ, Wang WT et al (2011) All-optical cascaded laser wakefield accelerator using ionization-induced injection. Phys Rev Lett 107:035001
17. Pollock BB et al (2011) Demonstration of a narrow energy spread, $\sim$0.5 GeV electron beam from a two-stage laser wakefield accelerator. Phys Rev Lett 107:045001
18. Gonsalves A, Nakamura K, Lin C et al (2011) Tunable laser plasma accelerator based on longitudinal density tailoring. Nat Phys 7(11):862–866
19. Geddes CGR, Nakamura K, Plateau GR et al (2008) Plasma-density-gradient injection of low absolute-momentum-spread electron bunches. Phys Rev Lett 100:215004
20. Davidson A, Tableman A, An W et al (2014) Implementation of a hybrid particle code with a PIC description in rz and a gridless description in ϕ into OSIRIS. arXiv:1403.6890
21. Vay JL (2007) Noninvariance of space- and time-scale ranges under a Lorentz transformation and the implications for the study of relativistic interactions. Phys Rev Lett 98:130405
22. Martins SF, Fonseca RA, Lu W et al (2010) Exploring laser-wakefield-accelerator regimes for near-term lasers using particle-in-cell simulation in Lorentz-boosted frames. Nat Phys 6(4):311–316
23. Lu W, Tzoufras M, Joshi C et al (2007) Generating multi-GeV electron bunches using single stage laser wakefield acceleration in a 3D nonlinear regime. Phys Rev ST Accel Beams 10:061301
24. Reicher S (1999) GENESIS 1.3: a fully 3D time-dependent FEL simulation code. Nucl Instrum Methods Phys Res Sect A 429(1–3):243–248
25. Tzoufras M, Lu W, Tsung FS et al (2008) Beam loading in the nonlinear Regime of plasma-based acceleration. Phys Rev Lett 101:145002
26. Tzoufras M, Lu W, Tsung FS et al (2009) Beam loading by electrons in nonlinear plasma wakes. Phys Plasmas 16(5):056705

Chapter 5
Numerical Instability Due to Relativistic Plasma Drift in EM-PIC Simulations

5.1 Introduction

Plasma based accelerators offer great potential for constructing compact accelerators that have numerous applications. The physical process in plasma based acceleration involves small spatial scale ($\sim \mu$m) and ultra-fast time scale ($\sim$ fs), and they are very difficult to diagnose directly in experiments. Therefore high fidelity numerical simulations play a critical role in the development of plasma based accelerators. With the rapid progress of plasma based accelerators, peoples begin to show great interests of high energy (~ 10 GeV or higher) and high quality electrons acceleration as drivers for compact light sources and the building blocks for a future linear collider. Due to the large variance of spatial and time scales involved in this physical process, full-scale 3D simulations of high energy plasma based accelerators become extremely time-consuming and challenging even with the best computer available today. We take a 10 GeV laser driven plasma wakefield accelerator with density $n_p = 10^{17}$ cm^{-3} as an example. In the laser propagation direction, the laser wavelength ($\sim 1 \mu$m) needs to be resolved. The plasma length is on the order of 0.1 m, which is five orders of magnitude larger than the laser wavelength. In the two transverse directions, the size of the accelerated electron beam can be much smaller compared with the transverse wake size, therefore at least a few hundred grids are needs to resolve the transverse motion of the electron beam. For such a simulation at lest 1 million cpu-hour is needed on today's super computers. Further more, if the accelerated beam has a very high quality (i.e., $\epsilon_n < 100$ nm), much more grids are needs in the transverse directions to resolve the very small size of the beam ($\sigma_r \propto \sqrt{\epsilon_n}$). In general, it is very expensive to study the high energy high quality laser plasma wakefield accelerators with full-dimensional PIC simulations.

Recently, boosted frame simulation [1, 2] was proposed to significantly reduce the computation cost of laser wakefield accelerators. In a frame moving with the laser, the laser wavelength increases and the total plasma length contracts [3], and these two factors can lead to a much reduced computing time. However, a numerical instability in the boosted frame simulations [4, 5] can occur easier to significantly degrade

© Springer Nature Singapore Pte Ltd. 2020

X. Xu, *Phase Space Dynamics in Plasma Based Wakefield Acceleration*,
Springer Theses, https://doi.org/10.1007/978-981-15-2381-6_5

the fidelity of the simulation, quite often contaminate the results meaningless. This instability originates from the relativistic drifting plasma in the moving frame. In this chapter, we analyze this numerical instability in details theoretically. In Sect. 5.2 we derive the numerical dispersion relation for cold plasma drift in the grids. The mode and the growth rate of the instability for the Yee solver are studied in Sect. 5.3. An asymptotic expression of the growth rate is given in Sect. 5.4. In Sect. 5.5 we adopt a spectral solver to reduce the growth of the instability. Finally a summary is provided in Sect. 5.6.

5.1.1 The Boosted Frame Simulations of LWFA

The idea of considering a physical system in another frame to simplify the problem is ubiquitous in physics. In 2007, J.-L. Vay pointed out that there may be a preferred frame of reference which minimizes the range of space and time scales for systems exhibiting large disparity of space and time scales [1]. Similar idea was also mentioned in a earlier proposal for numerical simulation of laser wakefield acceleration in 1990s [6]. In laser plasma wakefield acceleration, the smallest relevant space scale is determined by the laser wavelength, and the longest spatial scale is total plasma length which is on the order of $0.1 \sim 1$ m. This disparity between plasma length and laser wavelength can be reduced if the simulation is carried out in a frame co-moving with the laser pulse. In a frame moving with γ_b in the laser propagating direction, the laser wavelength (hence the cell size) increases by a factor of $(1 + \beta_b)\gamma_b$ and the plasma length contracts by a factor of γ_b by the Lorentz transformation. Therefore the total simulation step are reduced by a factor of $(1 + \beta_b)\gamma_b^2$ if the number of cells remains the same.

Recently, S. F. Martins et al. first implemented boosted frame simulation for laser wakefield acceleration and also used to explore several laser wakefield acceleration regimes. They showed that self-injected 11 GeV and externally injected 40 GeV electron beams can be obtained by using 1–3 PW lasers.

5.1.2 Numerical Noise Induced by Relativistic Plasma Drift in PIC Codes

The numerical noise is a critical issue in the boosted frame simulations. This noise can contaminate the interesting physics quickly. Martins et al. [4] and Vay et al. [5] have tried many schemes to suppress the noise, such as using high-order particle shapes and/or strong filters. They also found that a specific time step can effectively suppress the numerical noise for finite difference solvers, and named it as the "magical" time step [5]. This numerical noise is an instability originated from the relativistically drifting plasma named the numerical cherenkov instability (NCI).

Godfrey et al. studied the NCI induced by a plasma drifting in one dimension [7] and multi-dimensions [8]. Their analysis did not include relativistic mass effects in the plasma response (and therefore did not include relativistic drifts) and were applied to a code that solved the scalar and vector potential using the Coulomb gauge. However, most present day codes solve directly for the electric and magnetic fields and use a rigorous charge conserving current deposition [9]. Recent results indicate that an instability only occurs in 2D and 3D [4, 5].

To better understand and mitigate the observed instability, we present here an analysis of the numerical instability for a plasma drifting relativistically in multi-dimensions in which the electric field $\mathbf{E}$ and magnetic field $\mathbf{B}$ are solved for directly. We follow the basic method and notation of Ref. [8] and concentrate on situations where the plasma is cold but is drifting near the speed of light. The dispersion relation can be applied for various Maxwell Equation solvers including finite difference solvers and spectral solvers, and it includes finite size particle and aliasing effects. The dispersion relation predicts the growth rates and pattern of unstable modes in Fourier space. By comparing the theoretical predictions against the simulation results using an EM-PIC FDTD code OSIRIS [10] and the UCLA PIC Framework UPIC-EMMC [11] which is based on spectral solver, we conclude that the observed instability is indeed induced by the relativistic plasma drift. It is found that the unstable modes arise from the coupling of the Langmuir modes and electromagnetic modes in the drifting plasma. The coupling can occur in the fundamental Brillouin zones when EM waves with phase velocities less than the speed of light exist and from aliasing Langmuir modes and EM modes. The unstable modes are neither purely transverse nor longitudinal. We use the fact that the coupling in ω and $\mathbf{k}$ space are strongest between the Langmuir waves (including aliases) and EM waves to develop asymptotic expressions for the instability growth rate. These observations can then be used as a guide for selecting alternative Maxwell Equation solvers and smoothing schemes to mitigate the instability. Specifically, we show that how to eliminate the NCI for a spectral solver.

5.2 Numerical Dispersion Relation for Cold Plasma Drift

In this section we derive the dispersion relation of the system when a cold plasma drifts relativistically on the numerical grids. The effect from discrete Maxwell solvers is taken account while the motion of the particles is treated as continuous.

5.2.1 Derivation of Dispersion Relation

We mainly follow the notation in Ref. [8] to derive the numerical dispersion relation for a cold plasma drifting with relativistic velocities. We note that the 1D analysis [7] only predicts instability growth when sub-cycling is used (the particles are updated every N time steps of the field solver). The multi-dimensional analysis in [8] solves for scalar potential ϕ and vector potential $\mathbf{A}$ and is not valid for relativistic drifts. On the other hand, our analysis includes relativistic mass effects, is valid in multi-dimensions, allows for different types of field solvers and current deposition schemes. Since most EM-PIC codes now in use solve for the electric field $\mathbf{E}$ and magnetic field $\mathbf{B}$ directly (with finite difference or spectral solvers), we derive a numerical dispersion relation directly using these two quantities. Gaussian units will be used; in addition, particle mass and velocity will be normalized to electron mass and the speed of light.

For a multi-dimensional simulation setup in Cartesian coordinates, the EM field that is interpolated on a particle can be expressed as

$$\mathbf{E}(t, \mathbf{x}) = \sum_{m,\mathbf{n}} \overleftrightarrow{S_E}(t, m, \mathbf{x}, \mathbf{n})\mathbf{E}_{m,\mathbf{n}}$$

$$\mathbf{B}(t, \mathbf{x}) = \sum_{m,\mathbf{n}} \overleftrightarrow{S_B}(t, m, \mathbf{x}, \mathbf{n})\mathbf{B}_{m,\mathbf{n}} \tag{5.1}$$

where m is the time index and $\mathbf{n}$ is the grid index; $\overleftrightarrow{S}$ are the interpolation tensors used to obtain the appropriate field at $\mathbf{x}$ and $t = m\Delta t$; $\mathbf{E}_{m,\mathbf{n}}$ and $\mathbf{B}_{m,\mathbf{n}}$ stands for the electromagnetic forces at time grid index m and space grid index $\mathbf{n}$. For momentum conserving field interpolation $\overleftrightarrow{S_E}$ and $\overleftrightarrow{S_B}$ are equal and are scalar functions times the unit tensor while for energy conserving field interpolation $\overleftrightarrow{S_E}$ and $\overleftrightarrow{S_B}$ are not equal in each direction (the interpolation tensors for $\mathbf{E}$, $\mathbf{B}$, and $\mathbf{j}$ are given in Appendix B). The momentum change of the particle is related to the change in the distribution function of the plasma by the linearized Vlasov equation

$$\frac{\partial}{\partial t}f(t, \mathbf{x}, \mathbf{p}) + \frac{\mathbf{p}}{\gamma} \cdot \frac{\partial}{\partial \mathbf{x}}f(t, \mathbf{x}, \mathbf{p}) + q\left\{\mathbf{E}(t, \mathbf{x}) + \frac{\mathbf{p}}{\gamma} \times \mathbf{B}(t, \mathbf{x})\right\} \cdot \frac{\partial f_0}{\partial \mathbf{p}} = 0$$

where $\mathbf{p}$ is the particle momentum, and γ is the particle Lorentz factor. After Fourier transforming, the Vlasov equation becomes

$$f(\omega, \mathbf{k}, \mathbf{p}) = -iq\left\{\overleftrightarrow{S_E}(\omega, \mathbf{k})\mathbf{E}(\omega, \mathbf{k}) + \frac{\mathbf{p}}{\gamma} \times \{\overleftrightarrow{S_B}(\omega, \mathbf{k})\mathbf{B}(\omega, \mathbf{k})\}\right\} \cdot \frac{\partial f_0}{\partial \mathbf{p}}\left(\omega - \mathbf{k} \cdot \frac{\mathbf{p}}{\gamma}\right)^{-1} \tag{5.2}$$

Note that $\mathbf{E}$ and $\mathbf{B}$ are defined at the discrete grid position and discrete time step, so its Fourier transform in $(\omega, \mathbf{k})$ is periodic, i.e.,

$$\mathbf{E}(\omega, \mathbf{k}) = \mathbf{E}(\omega', \mathbf{k}')$$

$$\mathbf{B}(\omega, \mathbf{k}) = \mathbf{B}(\omega', \mathbf{k}') \tag{5.3}$$

where $\omega' = \omega + \mu\omega_g$, $k_i' = k_i + \nu_i k_{gi}$ and $\omega_g = 2\pi/\Delta t$, $k_{gi} = 2\pi/\Delta x_i$, $\mu = 0, \pm 1, \pm 2, \ldots, \nu_i = 0, \pm 1, \pm 2, \ldots$.

Note that when the EM fields are staggered (such as on a Yee mesh), there is an additional $(-1)^{\sum_i \nu_i}$ term for each component $\mathbf{E}(\omega', \mathbf{k}')$ and $\mathbf{B}(\omega', \mathbf{k}')$, where $\hat{i}$ is summed over the directions for which the specific component of the EM field is staggered a half-grid offset from where charge density is defined. We absorb these additional coefficients into the quantities $\overleftrightarrow{S_E}$ and $\overleftrightarrow{S_B}$ to keep Eq. 5.1 correct when $\mathbf{n}$ includes only integer indices.

Replacing $(w, \mathbf{k})$ with $(w', \mathbf{k}')$ in Eq. 5.2, we obtain

$$f(\omega', \mathbf{k}', \mathbf{p}) = -iq\left\{ \overleftrightarrow{S_E}(\omega', \mathbf{k}')\mathbf{E}(\omega, \mathbf{k}) + \frac{\mathbf{p}}{\gamma} \times \{\overleftrightarrow{S_B}(\omega', \mathbf{k}')\mathbf{B}(\omega, \mathbf{k})\} \right\} \cdot \frac{\partial f_0}{\partial \mathbf{p}} \left(\omega' - \mathbf{k}' \cdot \frac{\mathbf{p}}{\gamma}\right)^{-1}$$

$$(5.4)$$

The current density $\mathbf{j}$ due to the movement of the particles can be expressed as

$$\mathbf{j}(t, \mathbf{x}) = q\int \overleftrightarrow{S_j}(\mathbf{x}' - \mathbf{x})\frac{\mathbf{p}}{\gamma}f(\{m + 1/2\}\Delta t, \mathbf{x}', \mathbf{p})d\mathbf{x}'d\mathbf{p} \tag{5.5}$$

where $\overleftrightarrow{S_j}(\mathbf{x}' - \mathbf{x})$ is the tensor for the current deposit. After Fourier transforming we obtain

$$\mathbf{j}(\omega, \mathbf{k}) = q\sum_{\mu, \nu}(-1)^{\mu}\int \frac{\overleftrightarrow{S_j}(-\mathbf{k}')\mathbf{p}}{\gamma}f(\omega', \mathbf{k}', \mathbf{p})d\mathbf{p} \tag{5.6}$$

We can now proceed in the normal way to obtain a dispersion relation. We start from Faraday's and Ampere's Law,

$$\nabla \times \mathbf{E} = -\frac{\partial \mathbf{B}}{\partial t}$$

$$\nabla \times \mathbf{B} = \frac{\partial \mathbf{E}}{\partial t} + 4\pi\mathbf{j}$$

which upon Fourier transforming gives,

$$[\mathbf{k}]_E \times \mathbf{E} = [\omega]\mathbf{B} \tag{5.7}$$

$$[\mathbf{k}]_B \times \mathbf{B} = -[\omega]\mathbf{E} - 4\pi i\mathbf{j} \tag{5.8}$$

$[k]_E$ and $[k]_B$ are the finite difference operator of the corresponding Maxwell solver schemes for $\mathbf{E}$ and $\mathbf{B}$ fields (see Appendix B for details). We follow the notation in Ref. [8], and use $[\cdot]$ exclusively to indicate the finite difference operator. Applying $[\mathbf{k}]_B \times$ to both sides of Eq. 5.7, and using Eq. 5.8, we end up with the coupled wave equation for $\mathbf{E}$ and $\mathbf{j}$,

$$([\omega]^2 - [\mathbf{k}]_E \cdot [\mathbf{k}]_B + [\mathbf{k}]_E[\mathbf{k}]_B)\mathbf{E} = -4\pi i[\omega]\mathbf{j} \tag{5.9}$$

Using Eqs. 5.6 and 5.9, we could obtain

$$([\omega]^2 - [\mathbf{k}]_E \cdot [\mathbf{k}]_B + [\mathbf{k}]_E[\mathbf{k}]_B)\mathbf{E} = -4\pi i q \sum_{\mu,\nu}(-1)^{\mu}[\omega]\int \frac{\overleftrightarrow{S_j}\,(-\mathbf{k}')\mathbf{p}}{\gamma} f(\omega',\mathbf{k}',\mathbf{p})d\mathbf{p}$$

$$(5.10)$$

If we normalize the distribution function such that $f_0 = n_{p0} f_0^n$, use the definition of plasma frequency

$$\omega_p^2 = 4\pi q^2 n_{p0} \tag{5.11}$$

and use the expression for the distribution function in Eq. 5.4, we finally obtain

$$([\omega]^2 - [\mathbf{k}]_E \cdot [\mathbf{k}]_B + [\mathbf{k}]_E[\mathbf{k}]_B)\mathbf{E}$$

$$= -\omega_p^2 \sum_{\mu,\nu}(-1)^{\mu}\left\{\int \frac{\overleftrightarrow{S_j}\,(-\mathbf{k}')\mathbf{p}\,d\mathbf{p}}{\gamma\omega' - \mathbf{k}'\cdot\mathbf{p}}\left\{[\omega]\overleftrightarrow{S_E}(\omega',\mathbf{k}')\mathbf{E} + \frac{\mathbf{p}}{\gamma}\times\{\overleftrightarrow{S_B}(\omega',\mathbf{k}')([\mathbf{k}]_E\times\mathbf{E})\}\right\}\cdot\frac{\partial f_0}{\partial \mathbf{p}}\right\}$$

$$(5.12)$$

which is a generalized dispersion relation for a plasma of finite size particles drifting on a grid. We note that the use of additional smoothers and filters can be incorporated into the dispersion relation by adding additional $S_{SM}(\mathbf{k}')$ terms outside the summation over Brillouin zones (essentially it multiplies the ω_p^2 term).

5.2.2 Elements of Dispersion Relation Tensor

We next examine the dispersion relation in the limit of a cold plasma including the possibility that the drift is near the speed of light.

5.2.2.1 3D Case

Note that $\overleftrightarrow{S}$ for the fields and current has only three diagonal elements S_1, S_2, S_3 in each case. In 3D, we can expand Eq. 5.12 explicitly as

$$\begin{pmatrix} ([\omega]^2 - [k]_{E1}[k]_{B1} - [k]_{E2}[k]_{B2} - [k]_{E3}[k]_{B3})E_1 + [k]_{E1}[k]_{B1}E_1 + [k]_{E1}[k]_{B2}E_2 + [k]_{E1}[k]_{B3}E_3 \\ ([\omega]^2 - [k]_{E1}[k]_{B1} - [k]_{E2}[k]_{B2} - [k]_{E3}[k]_{B3})E_2 + [k]_{E2}[k]_{B1}E_1 + [k]_{E2}[k]_{B2}E_2 + [k]_{E2}[k]_{B3}E_3 \\ ([\omega]^2 - [k]_{E1}[k]_{B1} - [k]_{E2}[k]_{B2} - [k]_{E3}[k]_{B3})E_3 + [k]_{E3}[k]_{B1}E_1 + [k]_{E3}[k]_{B2}E_2 + [k]_{E3}[k]_{B3}E_3 \end{pmatrix}$$

$$= -\omega_p^2 \sum_{\mu,\nu}(-1)^{\mu}\int \frac{dp_1 dp_2 dp_3}{\gamma(\gamma\omega' - k_1'p_1 - k_2'p_2 - k_3'p_3)}\begin{pmatrix} S_{j1}p_1 \\ S_{j2}p_2 \\ S_{j3}p_3 \end{pmatrix}\begin{pmatrix} \partial f_0^n/\partial p_1 \\ \partial f_0^n/\partial p_2 \\ \partial f_0^n/\partial p_3 \end{pmatrix}^T \cdot$$

$$\begin{pmatrix} \gamma[\omega]S_{E1}E_1 + p_2 S_{B3}([k]_{E1}E_2 - [k]_{E2}E_1) + p_3 S_{B2}([k]_{E1}E_3 - [k]_{E3}E_1) \\ \gamma[\omega]S_{E2}E_2 + p_3 S_{B1}([k]_{E2}E_3 - [k]_{E3}E_2) + p_1 S_{B3}([k]_{E2}E_1 - [k]_{E1}E_2) \\ \gamma[\omega]S_{E3}E_3 + p_1 S_{B2}([k]_{E3}E_1 - [k]_{E1}E_3) + p_2 S_{B1}([k]_{E3}E_2 - [k]_{E2}E_3) \end{pmatrix} \tag{5.13}$$

This can be rewritten as

$$\overset{\leftrightarrow}{\epsilon}(\omega, k)\mathbf{E} = \begin{pmatrix} \epsilon_{11} & \epsilon_{12} & \epsilon_{13} \\ \epsilon_{21} & \epsilon_{22} & \epsilon_{23} \\ \epsilon_{31} & \epsilon_{32} & \epsilon_{33} \end{pmatrix} \begin{pmatrix} E_1 \\ E_2 \\ E_3 \end{pmatrix} = 0 \tag{5.14}$$

where we note that $\overset{\leftrightarrow}{\epsilon}$ is not the dielectric tensor. In addition, we are most interested in a cold plasma that is drifting. For such a case, the unperturbed normalized distribution function is given by

$$f_0^n = \delta(p_1 - p_0)\delta(p_2)\delta(p_3) \tag{5.15}$$

where $p_0 = \gamma_0 v_0$, v_0 is the drifting velocity of the plasma and γ_0 is the corresponding relativistic factor. Substituting the above form for f_0^n, Eq. 5.15, into Eq. 5.13, and carrying out the integration we obtain all the elements in the tensor as

$$\epsilon_{11} = [\omega]^2 - [k]_{E2}[k]_{B2} - [k]_{E3}[k]_{B3}$$
$$- \frac{\omega_p^2}{\gamma_0} \sum_{\mu,\nu} (-1)^\mu \frac{S_{j1}\{S_{E1}[\omega]\omega'/\gamma_0^2 + v_0^2(S_{B3}k_2'[k]_{E2} + S_{B2}k_3'[k]_{E3})\}}{(\omega' - k_1'v_0)^2}$$

$$\epsilon_{12} = [k]_{E1}[k]_{B2} - \frac{\omega_p^2}{\gamma_0} \sum_{\mu,\nu} (-1)^\mu \frac{S_{j1}v_0 k_2'(S_{E2}[\omega] - v_0 S_{B3}[k]_{E1})}{(\omega' - k_1'v_0)^2}$$

$$\epsilon_{13} = [k]_{E1}[k]_{B3} - \frac{\omega_p^2}{\gamma_0} \sum_{\mu,\nu} (-1)^\mu \frac{S_{j1}v_0 k_3'(S_{E3}[\omega] - v_0 S_{B2}[k]_{E1})}{(\omega' - k_1'v_0)^2}$$

$$\epsilon_{21} = [k]_{E2}[k]_{B1} - \frac{\omega_p^2}{\gamma_0} \sum_{\mu,\nu} (-1)^\mu \frac{v_0 S_{j2} S_{B3}[k]_{E2}}{\omega' - k_1'v_0}$$

$$\epsilon_{22} = [\omega]^2 - [k]_{E1}[k]_{B1} - [k]_{E3}[k]_{B3} - \frac{\omega_p^2}{\gamma_0} \sum_{\mu,\nu} (-1)^\mu \frac{S_{j2}(S_{E2}[\omega] - v_0 S_{B3}[k]_{E1})}{\omega' - k_1'v_0}$$

$$\epsilon_{23} = [k]_{E2}[k]_{B3}$$

$$\epsilon_{31} = [k]_{E3}[k]_{B1} - \frac{\omega_p^2}{\gamma_0} \sum_{\mu,\nu} (-1)^\mu \frac{v_0 S_{j3} S_{B2}[k]_{E3}}{\omega' - k_1'v_0}$$

$$\epsilon_{32} = [k]_{E3}[k]_{B2}$$

$$\epsilon_{33} = [\omega]^2 - [k]_{E1}[k]_{B1} - [k]_{E2}[k]_{B2} - \frac{\omega_p^2}{\gamma_0} \sum_{\mu,\nu} (-1)^\mu \frac{S_{j3}(S_{E3}[\omega] - v_0 S_{B2}[k]_{E1})}{\omega' - k_1'v_0}$$

$$\tag{5.16}$$

The dispersion relation is then finally obtained from the condition that

$$\mathrm{Det}(\overset{\leftrightarrow}{\epsilon}) = 0 \tag{5.17}$$

which is valid in any number of dimensions.

5.2.2.2 1D and 2D Case

Much can be learned from examining the 1D and 2D limits to the general dispersion relation. In 1D simulations all physical quantities only depend on one coordinate x_1, hence $[\mathbf{k}]$, $\mathbf{k}$, and $\mathbf{k}'$ only have the $\hat{1}$-component. It follows then the elements of $\overleftrightarrow{\epsilon}$ are

$$\epsilon_{11} = [\omega]^2 - \frac{\omega_p^2}{\gamma_0} \sum_{\mu,\nu} (-1)^\mu \frac{S_{j1} S_{E1}[\omega]\omega'/\gamma_0^2}{(\omega' - k_1' v_0)^2}$$

$$\epsilon_{22} = [\omega]^2 - [k]_{E1}[k]_{B1} - \frac{\omega_p^2}{\gamma_0} \sum_{\mu,\nu} (-1)^\mu \frac{S_{j2}(S_{E2}[\omega] - S_{B3}[k]_{E1}v_0)}{\omega' - k_1' v_0}$$

$$\epsilon_{33} = [\omega]^2 - [k]_{E1}[k]_{B1} - \frac{\omega_p^2}{\gamma_0} \sum_{\mu,\nu} (-1)^\mu \frac{S_{j3}(S_{E3}[\omega] - S_{B2}[k]_{E1}v_0)}{\omega' - k_1' v_0}$$

$$\epsilon_{12} = \epsilon_{13} = \epsilon_{21} = \epsilon_{23} = \epsilon_{31} = \epsilon_{32} = 0 \tag{5.18}$$

Using Eq. 5.17, the dispersion relation for the 1D case consists of three uncoupled modes,

$$\epsilon_{11} = 0 \qquad \epsilon_{22} = 0 \qquad \epsilon_{33} = 0 \tag{5.19}$$

where each mode corresponds to separate components of the electric fields E_1, E_2, and E_3 respectively. Each of these modes is numerically stable as long as Δt is sufficiently small. If we take the limit $\Delta t \to 0$, and $\Delta x \to 0$, then Eqs. 5.18 and 5.19 reduce to the dispersion relations in a real drifting plasma.

Similarly, the elements of $\overleftrightarrow{\epsilon}$ in the 2D limit can be written as

$$\epsilon_{11} = [\omega]^2 - [k]_{E2}[k]_{B2} - \frac{\omega_p^2}{\gamma_0} \sum_{\mu,\nu} (-1)^\mu \frac{S_{j1}(S_{E1}\omega'[\omega]/\gamma_0^2 + S_{B3}[k]_{E2}k_2'v_0^2)}{(\omega' - k_1' v_0)^2}$$

$$\epsilon_{12} = [k]_{E1}[k]_{B2} - \frac{\omega_p^2}{\gamma_0} \sum_{\mu,\nu} (-1)^\mu \frac{k_2'v_0 S_{j1}(S_{E2}[\omega] - S_{B3}v_0[k]_{E1})}{(\omega' - k_1' v_0)^2}$$

$$\epsilon_{21} = [k]_{E2}[k]_{B1} - \frac{\omega_p^2}{\gamma_0} \sum_{\mu,\nu} (-1)^\mu \frac{S_{j2}S_{B3}[k]_{E2}v_0}{\omega' - k_1' v_0}$$

$$\epsilon_{22} = [\omega]^2 - [k]_{E1}[k]_{B1} - \frac{\omega_p^2}{\gamma_0} \sum_{\mu,\nu} (-1)^\mu \frac{S_{j2}(S_{E2}[\omega] - S_{B3}[k]_{E1}v_0)}{\omega' - k_1' v_0}$$

$$\epsilon_{33} = [\omega]^2 - [k]_{E1}[k]_{B1} - [k]_{E2}[k]_{B2} - \frac{\omega_p^2}{\gamma_0} \sum_{\mu,\nu} (-1)^\mu \frac{S_{j3}(S_{E3}[\omega] - S_{B2}[k]_{E1}v_0)}{\omega' - k_1' v_0}$$

$$\epsilon_{13} = \epsilon_{23} = \epsilon_{31} = \epsilon_{32} = 0 \tag{5.20}$$

Using Eq. 5.17, we can obtain the dispersion relation for the 2D case

$$\epsilon_{11}\epsilon_{22} - \epsilon_{12}\epsilon_{21} = 0 \qquad \epsilon_{33} = 0 \tag{5.21}$$

Note that E_3 is de-coupled from the other two directions.

The numerical features of a particular simulation setup can now be investigated by solving the corresponding numerical dispersion relation. Due to the use of the finite space and time steps, these dispersion relations not only contain terms from the lowest order Brillouin zones ($\mu = 0$ and $v = 0$), but also the space aliasing (summation over v) and time aliasing (summation over μ) terms [12]. The elements of the interpolation tensors $\overset{\leftrightarrow}{S}$, and finite difference operators $[\cdot]$ come into the expression due to the finite difference treatments when depositing currents and EM fields, and when solving the Maxwell Equations. The modifications to the dispersion relation lead to numerical instability in the otherwise stable physical system [7, 8, 12, 13].

5.2.3 EM Modes, and Wave-Particle Resonance

Before we examine the unstable modes predicted by the dispersion relation, it is worth making some observations regarding the form of the elements of $\overset{\leftrightarrow}{\epsilon}$. As pointed out by Godfrey [7, 8], a key difference between the dispersion relation in a real system versus a PIC system is the presence of wave-particle (or beam) resonances in the fundamental and higher order Brillouin zones. We focus on the **k** near the beam resonance line

$$\omega' - k'_1 v_0 = 0. \tag{5.22}$$

The numerical solution of Eq. 5.17 for each mode can be analytically obtained by keeping only the corresponding μ and v terms in Eq. 5.16 since these terms are dominant near the corresponding resonance lines. Note for the NCI issues, we find it is a good approximation to truncate the sum of $v_{2,3}$ and only keep the $v_{2,3} = 0$ term. For simplicity, we consider the 2D case. After some algebra, the corresponding dispersion relation $\epsilon_{11}\epsilon_{22} - \epsilon_{12}\epsilon_{21} = 0$ can be written as

$$\left((\omega' - k'_1 v_0)^2 - \frac{\omega_p^2}{\gamma_0^3}(-1)^\mu \frac{S_{j1} S_{E1} \omega'}{[\omega]} \right) \times$$

$$\left([\omega]^2 - [k]_{E1}[k]_{B1} - [k]_{E2}[k]_{B2} - \frac{\omega_p^2}{\gamma_0}(-1)^\mu \frac{S_{j2}(S_{E2}[\omega] - S_{B3}[k]_{E1} v_0)}{\omega' - k'_1 v_0} \right) + C = 0 \tag{5.23}$$

where C is a coupling term in the dispersion relation

$$C = \frac{\omega_p^2}{\gamma_0} \frac{(-1)^\mu}{[\omega]} \left\{ S_{j1} S_{E1} \omega'[k]_{E2}[k]_{B2}(v_0^2 - 1) + S_{j2} S_{E2}[k]_{E2}[k]_{B2}(\omega' - k_1' v_0) \right.$$

$$\left. + S_{j1}[k]_{E2}(S_{E2}[k]_{B1} k_2 v_0 - S_{B3} k_2 v_0^2 [\omega]) \right\} \tag{5.24}$$

Much can be learned by investigating the new form of the dispersion relation Eq. 5.23. First, in the continuous limit, we have $[\omega] \to \omega$, $S_{E,B} \to 1$, so the coupling term C vanishes; second, the two factors in the first term of Eq. 5.23 are the Lorentz transformation of the dispersion relation of the Langmuir (longitudinal) mode, and the EM (transverse) mode in a stationary plasma, which in the continuous limit reduce to

$$(\omega - k_1 v_0)^2 - \frac{\omega_p^2}{\gamma_0^3} = 0$$

$$\omega^2 - k_1^2 - k_2^2 - \frac{\omega_p^2}{\gamma_0} = 0 \tag{5.25}$$

Consequently, we can identify the numerical Langmuir modes and EM modes for a drifting plasma as

$$(\omega' - k_1' v_0)^2 - \frac{\omega_p^2}{\gamma_0^3} (-1)^\mu \frac{S_{j1} S_{E1} \omega'}{[\omega]} = 0 \tag{5.26}$$

$$[\omega]^2 - [k]_{E1}[k]_{B1} - [k]_{E2}[k]_{B2} - \frac{\omega_p^2}{\gamma_0} (-1)^\mu \frac{S_{j2}(S_{E2}[\omega] - S_{B3}[k]_{E1} v_0)}{\omega' - k_1' v_0} = 0 \tag{5.27}$$

In addition, from Eq. 5.23 we see that when finite grid sizes and time steps are used neither Eq. 5.26 nor Eq. 5.27 leads to instability (if the Courant limit is satisfied). Therefore it becomes clear that the NCI is caused by the numerical coupling between modes which are purely longitudinal and purely transverse in the plasma rest frame due to the non-vanishing term C. Therefore reducing or eliminating the coupling term C is the key to mitigating the NCI. When the contributions of the plasma to the Langmuir modes and EM modes are small and can be neglected, the numerical Langmuir modes and EM modes reduce to the forms in the vacuum as

$$\omega' - k_1' v_0 = 0 \tag{5.28}$$

$$[\omega]^2 - [k]_{E1}[k]_{B1} - [k]_{E2}[k]_{B2} = 0 \tag{5.29}$$

In the following content, Eqs. 5.28 and 5.29 are frequently used to determine the positions of the numerical Langmuir modes and the EM modes. As a side note, it is evident that in 1D the coupling term vanishes, $C = 0$, in the numerical dispersion relation, hence no NCI is found in 1D.

Table 5.1 Crucial simulation parameters for the 2D relativistic plasma drift simulation. n_0 is the reference density, and $k_0 = \sqrt{4\pi q^2 n_0}/c$ is the corresponding wave number

Parameters	Values
Solver	Yee
Grid size $(k_0\Delta x_1, k_0\Delta x_2)$	$(0.1, 0.1)$
Time step $\omega_0\Delta t$	$0.9\times$ Courant limit
Boundary condition	Periodic
Simulation box size $(k_0 L_1, k_0 L_2)$	51.2×25.6
Plasma drifting Lorentz factor	$\gamma_0 = 50.0$
Plasma density	$n_{p0}/n_0 = 1$

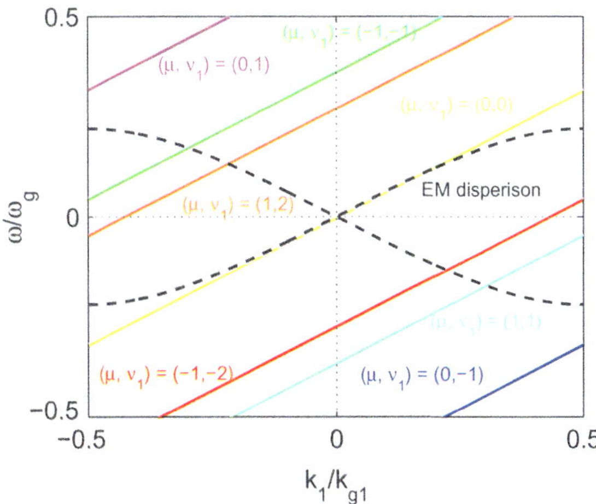

Fig. 5.1 The EM dispersion relation together with the beam resonance $\omega' - k_1'\beta = 0$ is shown. The parameters used to plot this figure are listed in Table 5.1

For each pair of (μ, ν_1) there is a corresponding Eq. 5.23. However, in PIC algorithm the range of (ω, k_1) for the quantities defined at discrete locations and time step is limited to the fundamental Brillouin zone $k_i \in (-k_{gi}/2, k_{gi}/2)$, $\omega \in (-\omega_g/2, \omega_g/2)$. As a result, not all the (μ, ν_1) wave-particle resonance line exist within the fundamental Brillouin zone. We can rewrite the wave-particle resonance line as

$$\hat{\omega} + \mu = \nu_0(\hat{k}_1 + \nu_1)\lambda_1 \tag{5.30}$$

where $\hat{\omega} = \omega/\omega_g$, $\hat{k}_i = k_i/k_{gi}$, $\lambda_i = \Delta t/\Delta x_i$. The criterion for the wave-particle resonance line to be inside the fundamental Brillouin zone are

$$|\nu_0\lambda_1\nu_1 - \mu| < 0.5 + 0.5\nu_0\lambda_1 \tag{5.31}$$

The NCI occurs where a resonance line intersects the EM dispersion relation. An example with the parameters in Table 5.1 are shown in Fig. 5.1, where we plot the resonance lines as solid lines and the EM dispersion relation in vacuum as dashed lines. Note for EM curves we only show $\hat{\omega}$ v.s. $\hat{k}_1$ at $\hat{k}_2 = 0$, but this line varies as $\hat{k}_2$ changes. For the NCI pattern and growth rates associated with each resonance line, we can numerically solve Eq. 5.23 using the corresponding (μ, ν_1). Note that we emphasize that for a particular resonance line, only one μ term in the elements of $\overleftrightarrow{\epsilon}$ is playing a dominant role.

5.3 Numerical Instability Induced by Relativistic Plasma Drift for the Yee Solver

Without loss of generality, we use the results in Sect. 5.2 to study the numerical instability induced by the relativistic plasma drift in a 2D system. According to the dispersion relation in 2D, we expect to observe instability in E_2 (and B_3). By calculating the maximum imaginary part of ω for real values of (k_1, k_2) for Eq. 5.21, we can obtain the characteristic pattern of the instability in Fourier space, as well as the growth rate of the instability. We can also plot the real part of ω for k_1, k_2. These results can be used later to compare with the simulation results.

The dispersion relation is general and can be used to examine different choices in Maxwell Equation solvers, differences between energy and momentum conserving field interpolation, differences between charge conserving and direct current deposition schemes, and the use of smoothing and low pass filters. In this section, we study the NCI for the Yee solver [14] as example. Other finite difference solvers, including the Karkkainen [15] and 4th-order solver [16] are discussed in Sect. 5.4.2. And the NCI and its elimination for the spectral solver are studied in Sect. 5.5 systematically.

5.3.1 Theoretical Analysis of the 2D Dispersion Relation

We illustrate the instability using a 2D case with the standard Yee solver [14]. We choose the grid parameters and time step that satisfies the Courant Condition [13] to eliminate known numerical instability from the EM modes. We use the parameters in Table 5.1, and substitute the finite difference operators for the Yee solver (see Appendix B) into the 2D dispersion relation. We assume linear (area) interpolation, momentum conserving field interpolation, and a charge conserving current deposition (see Appendix B).

After obtaining all the roots (ω, k_1, k_2), we plot the dependence of the growth rate in the (ω_r, k_1) space (Fig. 5.2c), as well as in the (k_1, k_2) space (Fig. 5.2d). It is evident that all the instabilities are near the main or aliased beam resonances. Since the terms with $|\mu| \leq 1$, and $|\nu_1| \leq 1$ are the most important, we neglect higher order terms when solving Eq. 5.21. Higher order μ and ν terms can be included in the summation if needed. These additional terms lead to additional unstable modes in (k_1, k_2) space with lower growth rates as well as to very small modifications to the

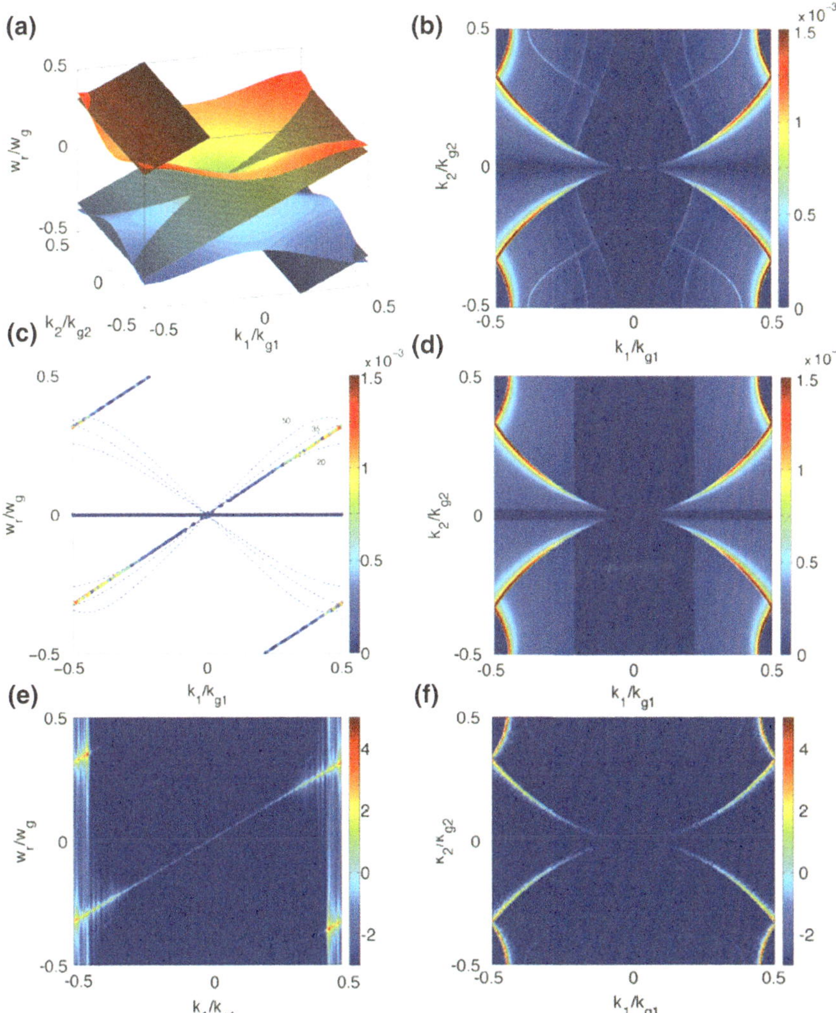

Fig. 5.2 Numerical instability pattern in the Yee solver. Growth rates are color-coded, and normalized with ω_g. **a** EM modes intersect with the main beam resonance ($\mu = 0$, $\nu = 0$), and first order space aliasing beam resonances ($\mu = 0$, $\nu_1 = \pm 1$); **b** is the instability pattern ($\mu = 0$, $|\nu_1| \leq 4$) in (k_1, k_2) space, plotted using Eqs. 5.38 and 5.39; **c** and **d** are the instability pattern ($|\mu| \leq 1$, $|\nu_1| \leq 1$) in (ω_r, k_1) and (k_1, k_2) spaces obtained from solving Eqs. 5.20 and 5.21. EM modes for different propagating angles [in degree] and the beam resonances are likewise plotted in (**c**). **e** presents the corresponding simulation results in (ω_r, k_1) space, and (**f**) in (k_1, k_2) space. Data in **e** and **f** show the modes present at $t = 100\,\omega_p^{-1}$, and are not a measurement of the growth rates (Reprinted from Ref. [17]. Copyright with kind permission from Elsevier)

growth rate and location of the original modes. A plot in (k_1, k_2) space with more terms included are presented in Fig. 5.2b, in which we use the asymptotic expression Eq. 5.41 for the growth rate (see Sect. 5.4 for more details).

While the results in Fig. 5.2c, d are numerically calculated from Eq. 5.21, the location of the unstable modes can also be conveniently predicted by plotting the intersection of the Langmuir modes Eq. 5.28 and EM modes Eq. 5.29 and in (k_1, k_2, ω_r) space. This is shown in a 3D plot (Fig. 5.2a). By examining the unstable pattern in (k_1, k_2) space we see that the central part of the pattern comes from the intersections of the EM modes and main Langmuir mode ($\mu = 0$ and $\nu = 0$), while the part at the four corners can be identified with the intersections of the EM modes and first order spatial aliasing Langmuir modes ($\mu = 0$ and $\nu_1 = \pm 1$). As we argue in Sect. 5.4, a key to mitigating the instability is to manipulate the instability pattern through a careful choice of the Maxwell Equation solver. Making a plot in (k_1, k_2, ω_r) of the intersection of the EM modes and beam resonances for various solvers becomes a useful method for examining where the unstable modes reside without having to solve the full dispersion relation.

5.3.2   Simulation Study of the Instability

To compare with the results in Sect. 5.3.1, we conducted simulation studies in the 2D system using the EM-PIC code OSIRIS [10]. In these simulations, a neutral plasma with both the ion and electrons drifting in x_1 at the same relativistic Lorentz factor of $\gamma_0 = 50.0$ is initialized throughout the entire simulation box. Periodic boundary conditions for fields and particles are used. Other crucial parameters for the simulation setup are identical to the theoretical study in Sect. 5.3.1.

As is shown in Fig. 5.3a, the total EM energy starts to grow violently as the plasma drifts relativistically. The exponential growth indicates that a numerical instability occurs. In addition, the EM field energy in E_2 and B_3 and that in E_3 and B_2 are shown separately. As predicted by the 2D dispersion relation the E_3 and B_2 modes are stable and do not grow. The pattern of E_2 at $t = 100\,\omega_p^{-1}$ is plotted in Fig. 5.2e, f, and good agreement for the location and relative amplitude of the unstable modes is obtained when compared against the theoretical prediction (Fig. 5.2c, d).

The EM energy grows with a lower rate after $t = 110\,\omega_p^{-1}$ (Fig. 5.3a). The plasma density in this regime is highly modulated by the EM fields. The first order perturbation in plasma electron density (Fig. 5.3b) shows a similar pattern as for E_2 (Fig. 5.3d), which confirms they are coupling in the system. Note that no exponential energy growth can be seen in the E_3 field (Fig. 5.3c).

From the simulation we find that for later times after the instability has evolved into a nonlinear state, the same pattern in (k_1, k_2) space as that of the linear regime still exists. This indicates that the instability will remain near the intersections of the EM modes and beam resonances and that both the linear and nonlinear growth can be mitigated through eliminating or controlling the intersections.

We also carried out a numerical investigation of the 1D dispersion relation Eqs. 5.18 and 5.19 using the same simulation parameters as in Table 5.1 (with the 1D Courant condition). This confirms that there is no numerical instability under these

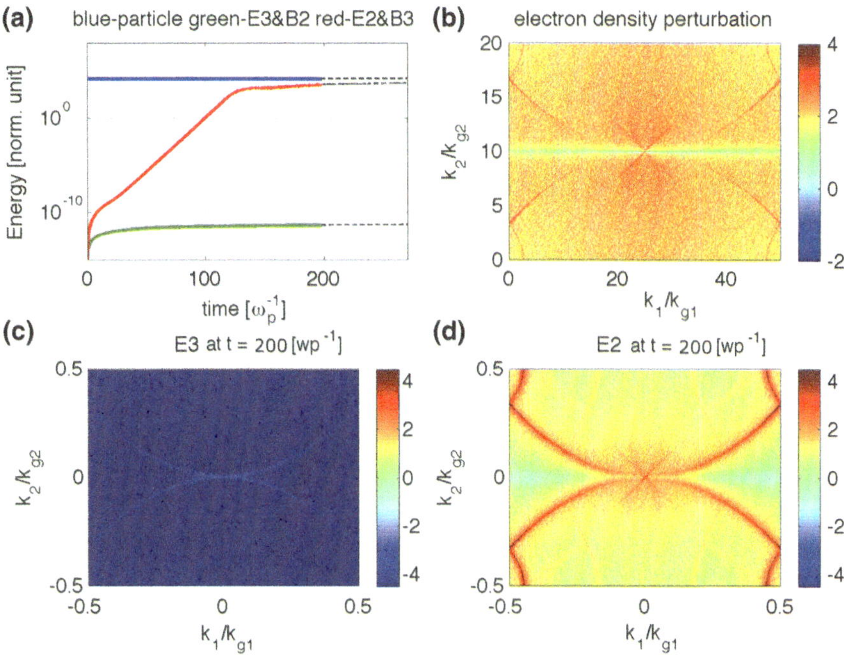

Fig. 5.3 We present in **a** the energy evolution of the EM energy for the two cases. The corresponding dotted line indicates their variation in time after $t = 200\ \omega_p^{-1}$; **b** is the plasma electron density perturbation in (k_1, k_2) space at $t = 200\ \omega_p^{-1}$. **c** Presents the E_3 in (k_1, k_2) space, and **d** Presents the E_2 in (k_1, k_2) space at $t = 200\ \omega_p^{-1}$ (Reprinted from Ref. [17]. Copyright with kind permission from Elsevier)

parameters which is expected since E_1 is de-coupled from E_2 and E_3 in Eq. 5.18 and each mode is itself stable.

5.4 Asymptotic Expression for Instability Growth Rate

In Sect. 5.3, we obtained the instability pattern and growth rate by numerically solving the dispersion relation equation, which is feasible on a modern laptop computer in 2D, but much more difficult in 3D. As mentioned above, the instabilities occur at the intersections of the beam resonances and EM modes. Taking advantage of this issue, we here derive an asymptotic expression for the solutions near the beam resonance $\omega' = k_1' v_0$. These expressions will not only speed up the instability pattern analysis in 3D, but also provide more insights into the dependence of instability pattern and growth rate to the grid sizes and time step used in simulation.

5.4.1 Derivation of Asymptotic Expression

We expand ω' around the beam resonance $\omega' = k_1' v_0$ as $\omega' = k_1' v_0 + \delta\omega'$, where $\delta\omega'$ is a small term. In addition, we will use the relativistic limit $v_0 \to 1$ and expand the finite operator $[\omega]$ as

$$[\omega] = \xi_0 + \delta\omega' \xi_1 \tag{5.32}$$

where

$$\xi_0 = \frac{\sin(\tilde{k}\Delta t/2)}{\Delta t/2} \qquad \xi_1 = \cos(\tilde{k}\Delta t/2) \tag{5.33}$$

and

$$\tilde{k} = k_1 + v_1 k_{g1} - \mu\omega_g \tag{5.34}$$

We will also truncate the summation over v_2 and v_3, keeping only the $v_2 = v_3 = 0$ terms. Around each beam resonance, only one pair of (μ, v_1) dominates the dispersion relation due to the $1/(\omega' - k_1' v_0)$ factor in the summation. Therefore, we can keep only one pair of (μ, v_1) when considering the instability around one given beam resonance. The elements of $\overleftrightarrow{\epsilon}$ can be written as

$$\epsilon_{11} = [\omega]^2 - [k]_{E2}[k]_{B2} - [k]_{E3}[k]_{B3} - \frac{Q_{11}}{\delta\omega'^2} \quad \epsilon_{12} = [k]_{E1}[k]_{B2} - \frac{Q_{12}}{\delta\omega'^2} \quad \epsilon_{13} = [k]_{E1}[k]_{B3} - \frac{Q_{13}}{\delta\omega'^2}$$

$$\epsilon_{21} = [k]_{E2}[k]_{B1} - \frac{Q_{21}}{\delta\omega'} \quad \epsilon_{22} = [\omega]^2 - [k]_{E1}[k]_{B1} - [k]_{E3}[k]_{B3} - \frac{Q_{22}}{\delta\omega'} \quad \epsilon_{23} = [k]_{E2}[k]_{B3}$$

$$\epsilon_{31} = [k]_{E3}[k]_{B1} - \frac{Q_{31}}{\delta\omega'} \quad \epsilon_{32} = [k]_{E3}[k]_{B2} \quad \epsilon_{33} = [\omega]^2 - [k]_{E1}[k]_{B1} - [k]_{E2}[k]_{B2} - \frac{Q_{33}}{\delta\omega'}$$

Using $\det(\overleftrightarrow{\epsilon}) = 0$, and dropping terms of higher order of $(\omega_p^2/\gamma)^2$, $(\omega_p^2/\gamma)^3$, …, we can obtain

$$A_1 \delta\omega'^2 + B_1 \delta\omega' + C_1 = 0 \tag{5.35}$$

where

$$A_1 = [\omega]^2 \left([\omega]^2 - [k]_{E1}[k]_{B1} - [k]_{E2}[k]_{B2} - [k]_{E3}[k]_{B3}\right)$$

$$B_1 = [k]_{E1}[k]_{B2}Q_{21} + [k]_{E1}[k]_{B3}Q_{31} - \left([\omega]^2 - [k]_{E2}[k]_{B2}\right)Q_{22} - \left([\omega]^2 - [k]_{E3}[k]_{B3}\right)Q_{33}$$

$$C_1 = -\left([\omega]^2 - [k]_{E1}[k]_{B1}\right)Q_{11} + [k]_{E2}[k]_{B1}Q_{12} + [k]_{E3}[k]_{B1}Q_{13} \tag{5.36}$$

Now we use the condition that (ω', k_1') sits near the EM modes,

$$[\omega]^2 \approx [k]_{E1}[k]_{B1} + [k]_{E2}[k]_{B2} + [k]_{E3}[k]_{B3} \tag{5.37}$$

Therefore we further expand $[\omega]$ to first order in A_1 since this term is sensitive near the EM mode, while keeping only the zero order of $[\omega]$ in B_1, and C_1. Note that this approximation is accurate for the instabilities occurring at the intersections of the beam resonances and the EM modes. For the highly localized instability discussed in Sect. 5.5 where the beam resonance and the EM mode do not intersect, we should keep to first order of $\delta\omega'$ in all terms. Now we obtain a cubic equation

$$A_2\delta\omega'^3 + B_2\delta\omega'^2 + C_2\delta\omega' + D_2 = 0 \tag{5.38}$$

with the coefficients

$$A_2 = 2\xi_0^3\xi_1$$
$$B_2 = \xi_0^2\left[\xi_0^2 - ([k]_{E1}[k]_{B1} + [k]_{E2}[k]_{B2} + [k]_{E3}[k]_{B3})\right]$$
$$C_2 \approx [k]_{E1}[k]_{B2}Q_{21} + [k]_{E1}[k]_{B3}Q_{31} - \left(\xi_0^2 - [k]_{E2}[k]_{B2}\right)Q_{22} - \left(\xi_0^2 - [k]_{E3}[k]_{B3}\right)Q_{33}$$
$$D_2 = -\left(\xi_0^2 - [k]_{E1}[k]_{B1}\right)Q_{11} + [k]_{E2}[k]_{B1}Q_{12} + [k]_{E3}[k]_{B1}Q_{13} \tag{5.39}$$

When the discriminant of this cubic equation

$$\Delta = 18A_2B_2C_2D_2 - 4B_2^3D_2 + B_2^2C_2^2 - 4A_2C_2 - 27A_2^2D_2^2 \tag{5.40}$$

satisfies the condition $\Delta < 0$, the cubic equation has one real root and two non-real complex conjugate roots. Therefore, by calculating the discriminant of the cubic equation Eq. 5.38, we can quickly identify the position of the instability for a particular (μ, ν_1). We can then use the general formula for the roots of a cubic equation to obtain the growth rate of the corresponding $\mathbf{k}$ mode. When calculating the instability growth rate, we obtain the imaginary part of roots $\Im\{\delta\omega'(\mathbf{k}, \mu, \nu_1)\}$ for each μ and ν_1 by solving Eqs. 5.38 and 5.39, and the growth rate $\Gamma(\mathbf{k}_0)$ for a particular mode $\mathbf{k}_0$ is chosen to be $\max\{\Im\{\delta\omega'(\mathbf{k}_0, \mu, \nu_1)\}\}$. Note that when solving for each $\Im\{\delta\omega'(\mathbf{k}, \mu, \nu_1)\}$, only the corresponding μ and ν_1 terms are kept in the above cubic equation. Equations 5.38 and 5.39 can be used to plot the growth rate in Fourier space, and can be conveniently simplified to 2D. In Fig. 5.2b we plot the asymptotic instability growth rate with $\mu = 0$, $|\nu_1| \leq 4$ for the 2D Yee solver using the same parameter listed in Table 5.1.

Near the transverse EM modes $[\omega]^2 \approx \xi_0^2 \approx [\mathbf{k}]_E \cdot [\mathbf{k}]_B$, we can drop the B_2 term in Eq. 5.38. According to our numerical results, we can further simplify the analytical expressions by dropping the small C_2 term. The asymptotic growth rate $\Gamma(\mathbf{k})$ for this mode corresponds to the maximum imaginary part for the three roots,

$$\Gamma(\mathbf{k}) \approx \frac{\sqrt{3}}{2}\left|\frac{\left(\xi_0^2 - [k]_{E1}[k]_{B1}\right)Q_{11} - [k]_{E2}[k]_{B1}Q_{12} - [k]_{E3}[k]_{B1}Q_{13}}{2\xi_0^3\xi_1}\right|^{1/3}$$

$$\approx \frac{\sqrt{3}}{2} \left| \frac{\omega_p^2 S_{j1} \{(S_{B3}\xi_0 - S_{E2}[k]_{B1})[k]_{E2}k_2 + (S_{B2}\xi_0 - S_{E3}[k]_{B1})[k]_{E3}k_3\}}{2\gamma_0 \xi_0^2 \xi_1} \right|^{1/3}$$

$$(5.41)$$

This expression shows the relation between the instability growth rate and grid sizes, time step, interpolation and smoothing functions, and finite difference operators. Note that from the positions of the interpolation functions we could immediately see that using higher particle shape, or using a stronger smoother helps mitigate the instability pattern at high $|\mathbf{k}|$ region, which agrees well with our simulation.

5.4.2 Parameter Scans for Minimal Instability Growth Rate

With the asymptotic expression Eqs. 5.38, 5.39, and also Eq. 5.41, we can greatly speed up the solution of numerical dispersion relation in 2D and 3D. In addition, the asymptotic expression makes the parameter scan to study the dependence of the instability pattern and growth rate between various grid sizes and time steps more convenient.

In Fig. 5.4, we scanned the grid sizes Δx_1 and time step $\Delta t / \Delta x_1$ for the 2D and 3D Yee solver, and Karkkainen solver, and compared the growth rates with the OSIRIS simulations. We have kept $\Delta x_1 = \Delta x_2 (= \Delta x_3)$ during the parameter scan for 2D (and 3D). We likewise plotted out the OSIRIS simulation data for $\Delta x_1 = 0.1$ together with the asymptotic data. There are several interesting points worth noting in Fig. 5.4. First, we can see there is an optimized time step [5] $\Delta t_m / \Delta x_1$ where the growth rate is minimized in most cases; on the other hand, the instability growth rate decreases monotonically as the grid sizes increase; second, when the grid sizes are square (2D) or cubic (3D), the optimized time step $\Delta t_m / \Delta x_1$ is an invariant for different Δx_1, in both the momentum conserving (MC) scheme, and energy conserving (EC) scheme; third, the instability growth rate for 2D and 3D are nearly the same for given Δx_1 and $\Delta t / \Delta x_1$ under the same field interpolation scheme; the values for the optimized time steps are also nearly the same in 2D and 3D (note that according to the asymptotic expression, the optimized time step for the Yee solver 3D EC scheme also resides at around $\Delta t_m / \Delta x_1 \approx 0.65$, but we did not plot it out since that Δt_m is beyond the Courant limit for this solver). The parameter scan using the asymptotic expression for the Karkkainen solver with the EC scheme shows the optimized time step at around $\Delta t_m / \Delta x_1 = 0.7$, which agrees with the results reported in Ref. [5]. However, according to our simulation and theoretical results, we found the optimized time step not only in the Karkkainen solver, but also in the Yee solver; and not only for the EC scheme, but also for the MC scheme. This is also reported in Ref. [18] for the 2D cases.

The fact that the optimized time step $\Delta t_m / \Delta x_1$ does not depend on Δx_1 for square (and cubic) cells for the Yee and Karkkainen solver is evident from Eq. 5.41. Applying the detailed form of finite difference operators $[k]_{Bi}$ for the Yee and Karkkainen

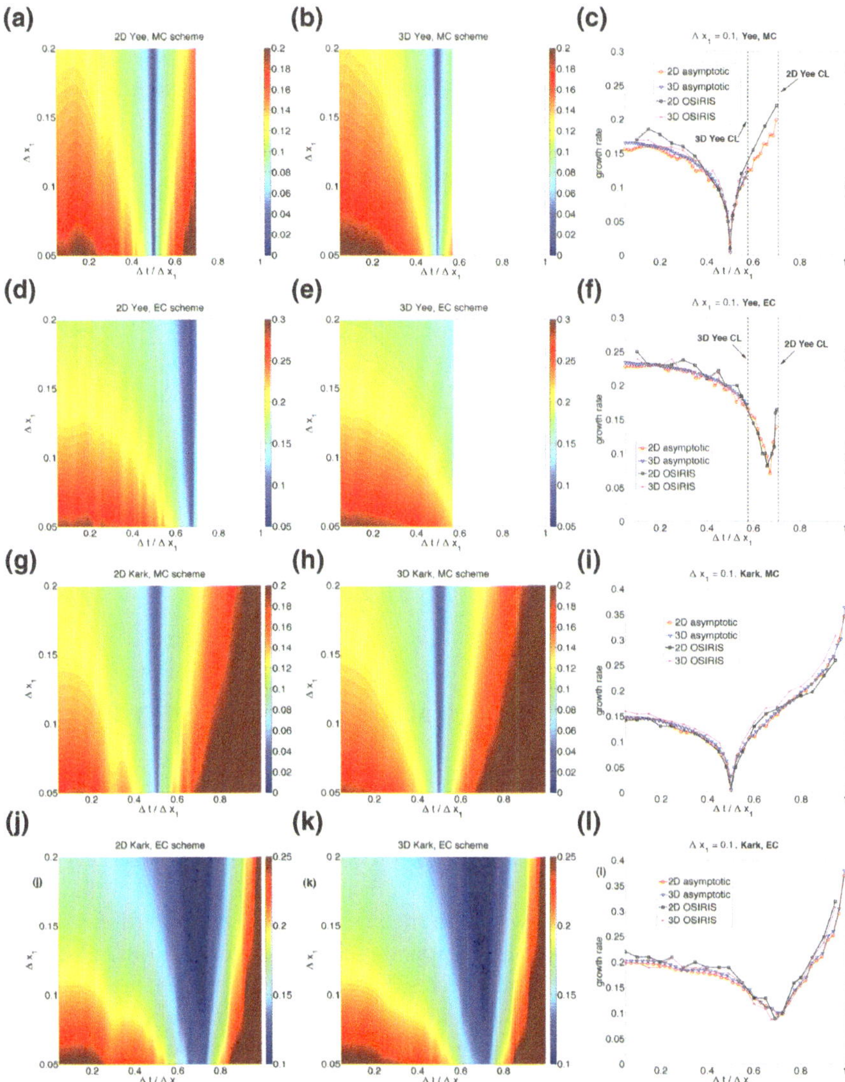

Fig. 5.4 Parameter scans of Δx_1 and $\Delta t/\Delta x_1$ for the Yee (first two rows), and Karkkainen (last two rows) solvers. Growth rates are color-coded in the first two columns. The first and third row uses momentum conserving (MC) scheme, while the second and fourth row uses the energy conserving (EC) scheme. The simulation results are likewise plotted in (**c**), (**f**), (**i**), and (**l**) at $\Delta x_1 = 0.1$ for comparisons. In **c** and **f** the dotted line at $\Delta t/\Delta x_1 \approx 0.577$ is the 3D Yee solver Courant limit (CL), and that at at $\Delta t/\Delta x_1 \approx 0.707$ is the 2D Yee solver CL. Similar lineout plots for 2D cases showing agreements between analytical and simulation results for an optimal time step were shown in Ref. [18] (Reprinted from Ref. [17]. Copyright with kind permission from Elsevier)

solvers (note they have the same $[k]_{Bi}$, see Appendix B), for both the MC and EC scheme, the expression of Γ can be expressed as $|k_1'|^{1/3}$ times a function of $\Delta t/\Delta x_1$, and k_i'/k_{gi}. (this function is different for different field interpolation schemes). Since k_i'/k_{gi} ranges from $(-0.5, 0.5)$ regardless of Δx_1 when calculating the growth rate, the extreme value of Γ resides at the same $\Delta t/\Delta x_1$ for different Δx_1 (although different field interpolation schemes give different optimized time steps). In particular, in the MC scheme the terms

$$S_{B3}\xi_0 - S_{E2}[k]_{B1} \tag{5.42}$$

and

$$S_{B2}\xi_0 - S_{E3}[k]_{B1} \tag{5.43}$$

(or only the first one in 2D) in Eq. 5.41 are zero when $\Delta t/\Delta x_1 = 1/2$ for these two solvers. As a result, both the Yee and Karkkainen solvers reach the minimal growth rate at $\Delta t/\Delta x_1 = 1/2$ in MC scheme, which agrees well with OSIRIS simulations. It is worth pointing out that, it is the staggering of the EM field in Yee mesh that leads to the minimal extreme value in Eq. 5.41.

5.5   Elimination of the Numerical Cerenkov Instability for Spectral EM-PIC Codes

As shown in the above sections, the NCI happens near the intersection of the Langmuir modes and the EM modes. Due to the fact that the EM dispersion curves for the finite difference solvers inevitably bend down at high $|k_1|$ such that the phase velocity of the electromagnetic waves on the grid is less than the drifting velocity of the charged particles, an instability is found in the $|\mathbf{k}|$ space as shown in Fig. 5.2. The physics we want to simulate always reside in the low $|\mathbf{k}|$ region, therefore it is difficult to mitigate the NCI due to the intersection of the main Langmuir mode and the EM mode while the NCI in the high $|\mathbf{k}|$ region due to the intersection of the first spatial aliasing of the Langmuir modes $(\mu, \nu_1) = (0, \pm 1)$ can be eliminated easily.

In this section, we explore the use of a spectral solver to eliminate the NCI. As we know, the phase velocity of the EM wave on the grid with a spectral solver is larger than the speed of light, therefore there is no intersection of the main Langmuir mode and the EM mode. The fastest growing instability occurs at the $(\mu, \nu_1) = (0, \pm 1)$, which is reside in the high $|\mathbf{k}|$ region and can be eliminated by low pass cutoff filter. When the fastest growing mode is eliminated a new mode which is highly localized in $\mathbf{k}$ space to four dots in 2D case and rings in 3D case can be seen. This new mode is also due to the coupling of the main Langmuir mode and EM mode. When the approximated main Langmuir mode Eq. 5.26 and approximated EM mode Eq. 5.27 are very close to each other, there can be an instability even the approximated expressions predict there is no intersection between these two modes. For time steps near the Courant limit, the growth rate of this high localized mode is usually one order of magnitude smaller than that of the fastest growing mode at the first spatial aliasing of the Langmuir modes. We found this new mode moves to higher $|\mathbf{k}|$ region outside the physically important region as the time step is reduced and can be also eliminated by low pass cutoff filter.

For the spectral solver, we will give a more rigorous theoretical analysis based on Eq. 5.23 which contains the new instability mode. Then, the pattern of the new mode and the fastest growing mode are studied through analytical formulas and simulations. At last, a LWFA simulations in the nonlinear regime are presented using the understanding of these NCI modes for the spectral solver.

5.5.1 *The NCI Modes for the Spectral Solver*

For the spectral solver,

$$[\mathbf{k}]_{E,B} = \mathbf{k}$$
$$S_{E1} = S_{E2} = S_{E3} \equiv S_E = S_l$$
$$S_{B1} = S_{B2} = S_{B3} \equiv S_B = \cos\frac{\omega\Delta t}{2}S_l.$$

where

$$S_l = \left(\frac{\sin(k_1\Delta x_1/2)}{k_1\Delta x_1/2}\right)^{l+1}\left(\frac{\sin(k_2\Delta x_2/2)}{k_1\Delta x_2/2}\right)^{l+1} \tag{5.44}$$

and l corresponds to the order of the particle shape. We can get the 3D form of Eq. 5.23 for the spectral solver as

$$\left([\omega]^2 - k_1^2 - k_2^2 - k_3^2 - \frac{\omega_p^2}{\gamma_0}(-1)^\mu S_j \frac{S_E[\omega] - S_B k_1 v_0}{\omega' - k_1' v_0}\right)$$
$$\left\{\left((\omega' - k_1' v_0)^2 - \frac{\omega_p^2}{\gamma_0^3}(-1)^\mu \frac{S_j S_E \omega'}{[\omega]}\right)\left([\omega]^2 - k_1^2 - k_2^2 - k_3^2 - \frac{\omega_p^2}{\gamma_0}(-1)^\mu S_j \frac{S_E[\omega] - S_B k_1 v_0}{\omega' - k_1' v_0}\right)\right.$$
$$\left. + \frac{\omega_p^2}{\gamma_0[\omega]}(-1)^\mu S_j(k_2^2 + k_3^2)\{v_0^2(S_E \omega' - S_B[\omega]) - v_0 v_1 S_E k_{g1}\}\right\} = 0 \tag{5.45}$$

For the instability mode near the resonance line, we can assume

$$\left([\omega]^2 - k_1^2 - k_2^2 - k_3^2 - \frac{\omega_p^2}{\gamma_0}(-1)^\mu S_j \frac{S_E[\omega] - S_B k_1 v_0}{\omega' - k_1' v_0}\right) \neq 0 \tag{5.46}$$
$$\left((\omega' - k_1' v_0)^2 - \frac{\omega_p^2}{\gamma_0^3}(-1)^\mu \frac{S_j S_E \omega'}{[\omega]}\right)\left([\omega]^2 - k_1^2 - k_2^2 - k_3^2 - \frac{\omega_p^2}{\gamma_0}(-1)^\mu S_j \frac{S_E[\omega] - S_B k_1 v_0}{\omega' - k_1' v_0}\right)$$

$$+ \frac{\omega_p^2}{\gamma_0[\omega]}(-1)^\mu S_j (k_2^2 + k_3^2)\{v_0^2(S_E \omega' - S_B[\omega]) - v_0 v_1 S_E k_{g1}\} = 0 \tag{5.47}$$

which, as in the 2D case, can be viewed as the coupling between the Langmuir and EM modes.

As same as in Sect. 5.4, we expand ω' around the beam resonance $\omega' = k_1' v_0$ as $\omega' = k_1' v_0 + \delta\omega'$, where $\delta\omega'$ is a small term. We keep to the first order of $\delta\omega'$ in all terms. In addition, we found it is sufficiently accurate if we neglect the ω^2/γ^3 term in the Langmuir mode in Eq. 5.47. This is why it is essentially the same to say that the instability occurs at wave-particle resonances, beam resonances, or at Langmuir resonances. Using these approximations, we obtain a cubic equation for $\delta\omega'$,

$$A_3 \delta\omega'^3 + B_3 \delta\omega'^2 + C_3 \delta\omega' + D_3 = 0 \tag{5.48}$$

where

$$A_3 = 2\xi_0^3 \xi_1$$

$$B_3 = \xi_0^2 \left[\xi_0^2 - k_1^2 - k_2^2 - k_3^2 - \frac{\omega_p^2}{\gamma}(-1)^\mu S_j \left(S_E \xi_1 - \kappa_1 S_B' k_1 \right) \right]$$

$$C_3 = \frac{\omega_p^2}{\gamma}(-1)^\mu S_j \left[\xi_0^2 (\kappa_0 S_B k_1 - S_E \xi_0) - \xi_1 (k_2^2 + k_3^2) S_E k_1 + \xi_0 (k_2^2 + k_3^2)(S_E - \kappa_1 S_B' \xi_0) \right]$$

$$D_3 = \frac{\omega_p^2}{\gamma}(-1)^\mu \xi_0 S_j (k_2^2 + k_3^2)(S_E k_1 - \kappa_0 S_B \xi_0) \tag{5.49}$$

where we separate the ω terms from S_B by writing $S_B = \cos(\omega \Delta t/2) S_B'$ and expand S_B to first order as

$$S_B = (\kappa_0 + \kappa_1 \delta\omega') S_B' \tag{5.50}$$

with $\kappa_0 = \cos(\tilde{k}_1 \Delta t/2)$ and $\kappa_1 = -\sin(\tilde{k}_1 \Delta t/2)\Delta/2$. The coefficients A_3 to D_3 are real, and completely determined by k_1 and k_2. As a result, by solving Eqs. 5.48 and 5.49 we can rapidly calculate the location and growth rate of the instability.

Following the prescription described above for each (μ, ν_1) modes, we systematically investigate the NCI modes for the spectral solver. In Fig. 5.5b, d, f we present the three sets of modes with the highest growth rate calculated by the analytical expressions Eqs. 5.48 and 5.49, for the parameters listed in Table 5.2, and for linear particle shapes ($l = 1$). Figure 5.5b shows the modes with $(\mu, \nu_1) = (0, \pm 1)$, which are the fastest growing NCI modes. Figure 5.5d shows the $(\mu, \nu_1) = (0, 0)$ modes, which have a highly localized pattern of four dots. Note that in Fig. 5.5d only one quadrant is plotted. These modes usually have a maximum growth rate one order of magnitude smaller than the $(\mu, \nu_1) = (0, \pm 1)$ modes. For the parameters listed in Table 5.2, the next fastest growing modes are the $(\mu, \nu_1) = (\pm 1, \pm 2)$ modes which have a maximum growth rate approximately 3 times smaller than the $(\mu, \nu_1) = (0, 0)$ modes for linear particle shape.

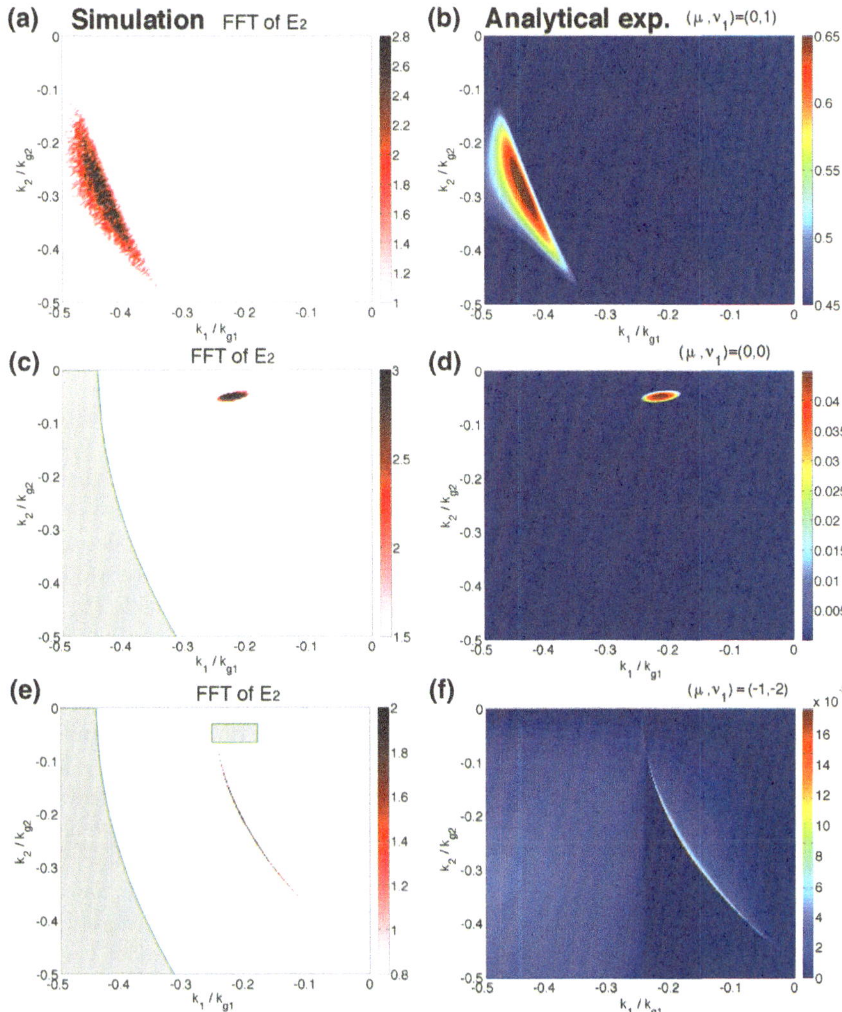

Fig. 5.5 **a**, **c**, and **e** are the FFT of E_2 in the 2D simulations using the parameters listed in Table 5.1. The filter applied in order to observe these modes are illustrated by the grey areas in the plots. **b**, **d**, and **f** are the corresponding predictions by using the expression Eqs. 5.48 and 5.49 (Reprinted from Ref. [19]. Copyright with kind permission from Elsevier)

We have similarly performed UPIC-EMMA simulations in 2D to observe various NCI modes in the spectral solver, and to compare with the theory presented above. The simulations use a neutral plasma drifting at relativistic velocity, with the Lorentz factor $\gamma_0 = 50.0$. The plasma has a uniform initial spatial distribution, and we used the parameters listed in Table 5.2. Note these parameters are those commonly used

Table 5.2 Crucial simulation parameters for the 2D relativistic plasma drift simulation. n_0 is the reference density, and $k_0 = \sqrt{4\pi q^2 n_0}/c$ is the corresponding wave number

Parameters	Values
Grid size $(k_0 \Delta x_1, k_0 \Delta x_2)$	$(0.2, 0.2)$
Time step Δt	$0.4 \Delta x_1$
Boundary condition	Periodic
Simulation box size $(k_0 L_1, k_0 L_2)$	102.4×102.4
Plasma drifting Lorentz factor	$\gamma_0 = 50.0$
Plasma density	$n_{p0}/n_0 = 100.0$

in the LWFA simulation in the Lorentz boosted frame, and the plasma density is 100 times larger than that used in Table 5.1.

Figure 5.5a, c, e show the simulation data of the FFT of E_2 at a particular time during the exponential EM energy growth from the NCI. Figure 5.5a shows results from a simulation with no low-pass filter, and the most prominent modes are the $(\mu, \nu_1) = (0, \pm 1)$ modes. To generate the frames in the middle row, we use a low-pass filter to eliminate the $(\mu, \nu_1) = (0, \pm 1)$ modes. This makes the unstable $(\mu, \nu_1) = (0, 0)$ modes more noticeable. It is shown in Fig. 5.5c that the $(\mu, \nu_1) = (0, 0)$ modes have a highly localized pattern of four dots, which agrees with the prediction of the analytic expression. According to Eq. 5.31, there is no intersection between $(\mu, \nu_1) = (1, 1)$ resonance and $(\mu, \nu_1) = (-1, -1)$ resonance and the EM dispersion relation, so the next set of modes of interest are the $(\mu, \nu_1) = (1, 2)$ and $(\mu, \nu_1) = (-1, -2)$ modes. To make the $(\mu, \nu_1) = (\pm 1, \pm 2)$ mode more noticeable, we use a low-pass filter to filter out the $(\mu, \nu_1) = (0, \pm 1)$ mode, plus a four-dot mask filter to remove the $(\mu, \nu_1) = (0, 0)$ mode. As shown in Fig. 5.5e, f, the locations of these modes in the simulation agree with the analytic prediction. As a side note, this numerical experiment also shows the simplicity and flexibility of using filters (masks) with complicated shapes in a spectral EM-PIC code to control the unphysical NCI growth.

5.5.2 The Positions and the Growth Rates of the NCI Modes for the Spectral Solver

According to both the theory and simulations, in the parameter space we are interested in, we usually categorize the NCI for a spectral solver into three categories: the fastest growing modes at $(\mu, \nu_1) = (0, \pm 1)$; the second fastest growing modes at $(\mu, \nu_1) = (0, 0)$; and higher order NCI modes with $|\nu_1| > 1$ that have an even smaller growth rate. In the following we will discuss how the locations and positions of these modes change with the simulation parameters.

For the NCI modes with $|\nu_1| \geq 1$, the instability resides around the intersections of the Langmuir mode and EM mode:

$$\left[\frac{\sin\left(k_1 v_0 \Delta t/2 + \xi \Delta t/2\right)}{\Delta t/2}\right]^2 = k_1^2 + k_2^2 + k_3^2 \tag{5.51}$$

where $\xi = v_0 v_1 k_{g1} - \mu \omega_g$. If we use the normalization as $\hat{\xi} = \xi/\omega_g$, $\hat{k}_1 = k_1/k_{g1}$, $\hat{k}_2 = k_2/k_{g2}$, $\hat{k}_3 = k_3/k_{g3}$ the equation above can be written as

$$\sin^2\left(\pi \hat{k}_1 v_0 \lambda_1 + \pi \hat{\xi}\right) = \pi^2 \lambda_1^2 \left(\hat{k}_1^2 + \hat{k}_2^2 + \hat{k}_3^2\right) \tag{5.52}$$

where $\hat{\xi} = v_0 v_1 \lambda_1 - \mu$ and square cells are used. The positions of the unstable NCI modes in $\hat{\mathbf{k}}$ space depends only on Δt and Δx_i through their ratio λ_1. Therefore, the position of the $|v_1| \geq 1$ NCI does not change if one keeps the ratio of time step to cell size.

For the NCI at $(\mu, v_1) = (0, 0)$, there is no intersection between the corresponding fundamental Langmuir mode and the EM mode (as can be seen by plotting Eqs. 5.26 and 5.27 in (ω, k_1) space, see Fig. 5.6a). The two modes interact at highly localized positions determined by the coupling term in Eq. 5.23. To show how the coupling term C modifies the Langmuir and EM modes, we solve Eq. 5.47 both with and without the coupling term by numerically forcing the coupling term to be zero and plot the solutions in Fig. 5.6. The parameters used in solving Eq. 5.47 numerically are the same as in Table 5.2, with $(\mu, v_1) = (0, 0)$. It is evident in Fig. 5.6a, c that when the coupling term is present, the fundamental Langmuir mode and EM mode are coupled near $-0.057 \leq \hat{k}_2 \leq -0.037$. In Fig. 5.6c where the growth rate is plotted, it becomes clear that in this range of k_2 where the fundamental Langmuir mode and EM mode are coupled, the two modes become complex conjugate pairs with one of them corresponding to instability in this range of k_2. In Fig. 5.6b, d, we scan ranges in both k_1 and k_2.

We next investigate the sensitivity of the growth rate and location in $\mathbf{k}$ space to the simulation parameters for the NCI at the fundamental mode $(\mu, v_1) = (0, 0)$. Note that we define the position of these modes at the value of $(\hat{k}_1, \hat{k}_2)$ where the growth rate is maximum. In reality, there is a range (although highly localized) of modes that go unstable. Figure 5.7a–d shows how the positions and growth rates of those modes change with plasma density and time step. For each simulation setup we plot both the simulation results and the predictions from the analytical expressions. When changing the grid sizes we fix $\Delta x_1 = \Delta x_2$. Figure 5.7a shows that when the grid sizes increases, the position of the $(\mu, v_1) = (0, 0)$ NCI moves farther away from the center of the (k_1, k_2) plot where the interesting real physics resides (red curve in Fig. 5.7a). We keep Δt constant as Δx_1 changes. This mode also moves farther away from the interesting physics when the time step decreases (see red curve in Fig. 5.7b). Furthermore, as shown in Fig. 5.7c, the growth rate decreases as the time step decreases [blue curve], which is not the case for the fastest growing modes of the NCI. The growth rate also decreases when the grid size increases while keeping Δt fixed (Fig. 5.7c red curve). When the density of the plasma increases (while fixing

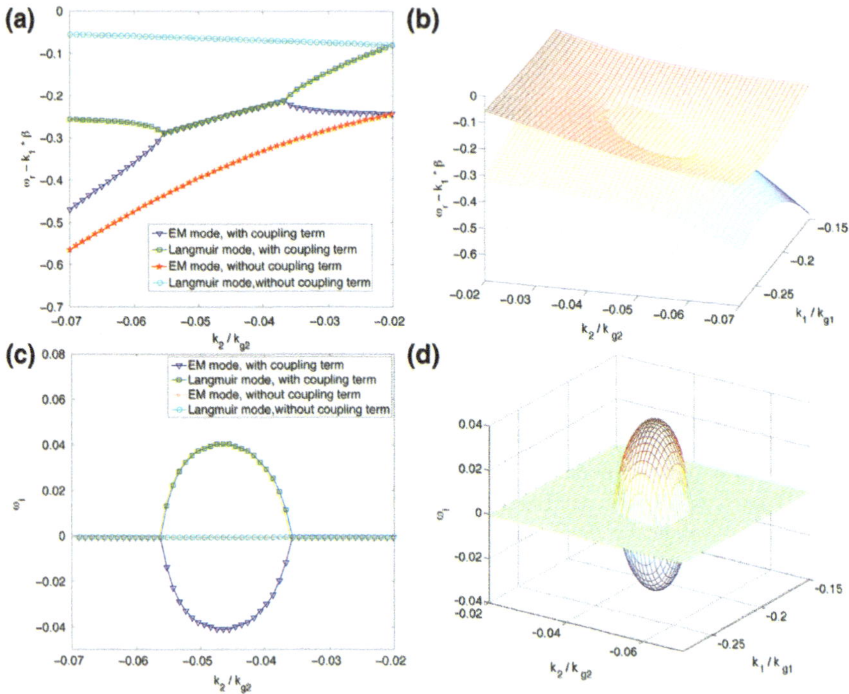

Fig. 5.6 Roots of Eq. (5.23) under the parameters listed in Table 5.1. **a** and **c** shows the real, and imaginary parts of the roots between $\hat{k}_1 = 0.21$, and $-0.07 \leq \hat{k}_2 \leq -0.2$, both with and without the coupling terms; meanwhile **b** and **d** shows the real and imaginary part of the roots in the range $-0.28 \leq \hat{k}_1 \leq -0.15$ and $-0.07 \leq \hat{k}_2 \leq -0.02$ (Reprinted from Ref. [19]. Copyright with kind permission from Elsevier)

$\gamma_0 = 50$), the position of the $(\mu, \nu_1) = (0, 0)$ NCI moves away from the center in (k_1, k_2) space, and the growth rate of the unstable modes increases (Fig. 5.7d).

A parameter scan which shows how the growth rate and position of the most unstable $(\mu, \nu_1) = (0, 0)$ modes change with different choices of the grid sizes and time step, is shown in Fig. 5.8. Note that we are keeping $\Delta x_1 = \Delta x_2$ in the parameter scan. By examining Fig. 5.8, we see that by reducing the $\Delta t / \Delta x_1$ ratio, the instability at the fundamental Langmuir mode moves farther towards larger $\hat{k}_1$ and the growth rate decreases. This indicates when the grid size is restricted to resolve the characteristic length of physical modes, the position of the $(\mu, \nu_1) = (0, 0)$ mode can be moved to larger $\hat{k}_1$ by simply using a smaller time step.

Meanwhile, when the time step is fixed, the growth rates of higher order NCI ($|\nu_1| > 1$) unstable modes can be efficiently reduced by using higher order particle shapes. In Fig. 5.9a we show how using different particle shapes changes the growth rate of the various NCI modes. We used the parameters in Table 5.2 for this figure. The result indicates that, while using higher order particle shapes is very efficient in reducing the growth rate of higher order NCI modes, it is less efficient for the $(\mu, \nu_1) = (0, 0)$ mode. We also compared results with different grid sizes (while

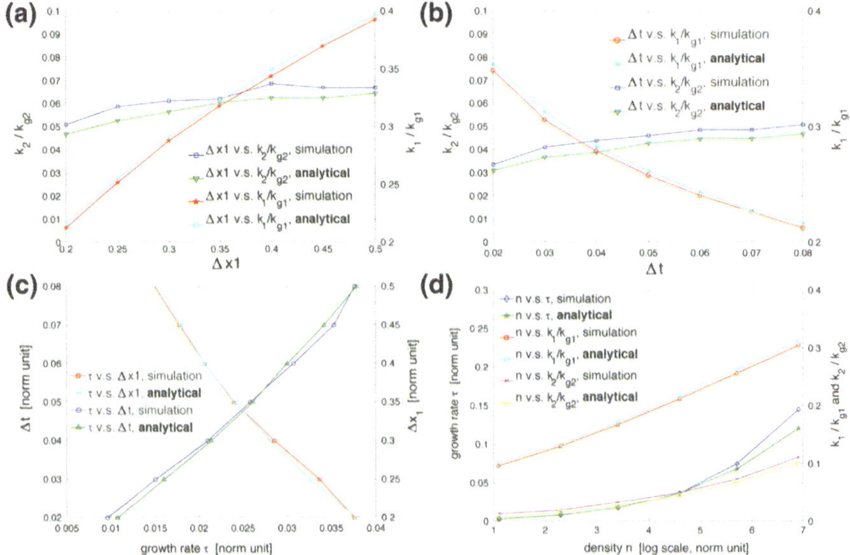

Fig. 5.7 Dependence of the position $(\hat{k}_1, \hat{k}_2)$, as well as the growth rate τ of the NCI at the fundamental Langmuir mode to grid sizes Δx_1 (with $\Delta x_1 = \Delta x_2$ fixed), time step Δt, and plasma density n_0 (Reprinted from Ref. [19]. Copyright with kind permission from Elsevier)

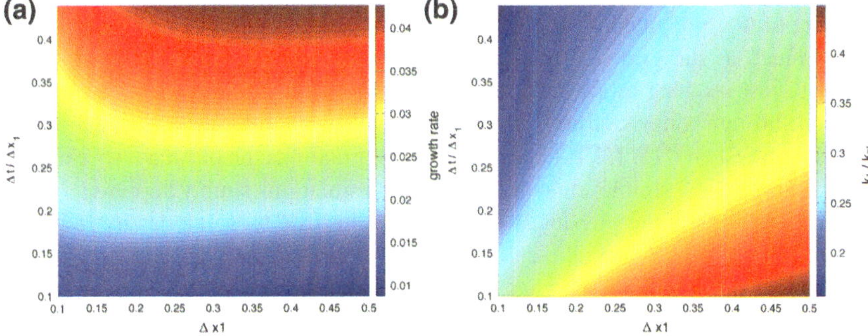

Fig. 5.8 Dependence of the **a** growth rate, and **b** k_1 position of NCI at the fundamental Langmuir mode for grid sizes $0.1 \leq \Delta x_1 \leq 0.5$ (with $\Delta x_1 = \Delta x_2$ fixed), and $0.1 \leq \Delta t/\Delta x_1 \leq 0.45$ (Reprinted from Ref. [19]. Copyright with kind permission from Elsevier)

fixing $\Delta t/\Delta x_1 = 0.4$), as shown in Fig. 5.9b. It indicates that reducing the grid size (while fixing $\Delta t/\Delta x_1$) helps reduce the growth rate of the $(\mu, \nu_1) = (0, 0)$ mode, but not for the modes with $\nu_1 \neq 0$.

Based on this understanding of the behavior of the unstable NCI modes, we now discuss approaches for controlling it. Once the NCI is adequately controlled, high fidelity simulations of relativistically drifting plasma can be carried out. We concentrate on spectral solvers and note that others are developing approaches for finite difference solvers [20]. The new form for the dispersion relation in Eq. 5.23

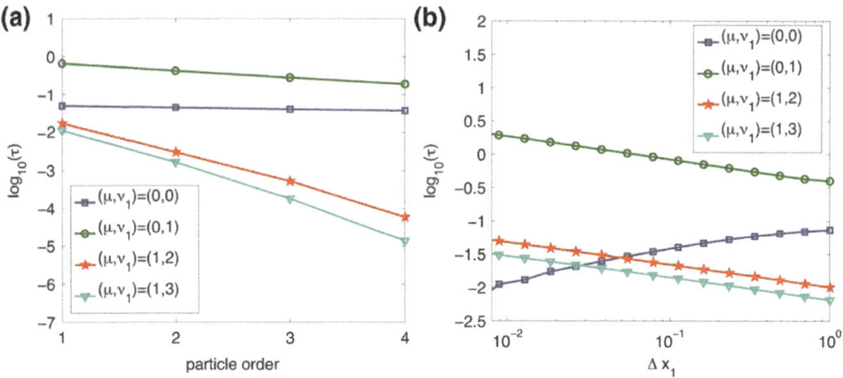

Fig. 5.9 **a** shows the dependence of the growth rate on grid size for various ν_1 modes; **b** shows the dependence of the growth rate τ on particle shapes. Parameters listed in Table 5.1 are used for these plots (Reprinted from Ref. [19]. Copyright with kind permission from Elsevier)

can also be used to investigate the NCI for other solvers and we leave this for future work.

The approach we use is to first move the unstable modes to large $\hat{\mathbf{k}}$'s that are outside the region $\hat{\mathbf{k}}$ where important physics is occurring. As discussed earlier, for the spectral solver the fastest growing modes at $(\mu, \nu_1) = (0, \pm 1)$ exist at large $|\hat{\mathbf{k}}|$ (the edge of the fundamental Brillouin zone), and their location in $\hat{\mathbf{k}}$ space does not change much as the grid size and time step are varied. As shown in Fig. 5.5, the second fastest growing mode at $(\mu, \nu_1) = (0, 0)$ is localized in $\hat{\mathbf{k}}$-space and can be removed through a mask (filter). However, these modes may exist near modes of physical interest, and for LWFA boosted frame simulations the plasma only exists in a small region of the simulation window. For such situations simply applying a mask filter may also effect the physics. We therefore eliminate those modes by reducing the time step (while keeping the cell size fixed). As shown in Fig. 5.8a, b, this both moves the unstable modes to higher $\hat{k}_1$ and lowers the growth rate. Once the modes are outside the range of physical modes a filter (mask) can be applied if necessary.

To investigate how reducing the time step changes the NCI, 2D simulations using the same parameters as the one shown in Fig. 5.5, but with a reduced time step of $\Delta t = 0.1 \Delta x_1$ are conducted. The corresponding beam resonances for this time step are illustrated in Fig. 5.10a, while the corresponding simulation data and analytical prediction for $\Delta t = 0.1 \Delta x_1$ are shown in Fig. 5.10c–f. From Fig. 5.10c, e we see as expected that when the time step is reduced, the growth rate and pattern of the fastest growing modes at $(\mu, \nu_1) = (0, \pm 1)$ do not change much (compared with Fig. 5.5a, b). However, for the $(\mu, \nu_1) = (0, 0)$ modes shown in Fig. 5.10d, f, the locations move away from the center (compared with Fig. 5.5c, d), while the growth rate is reduced by approximately a factor of 4.

In addition, when the time step is reduced to suppress the $(\mu, \nu_1) = (0, 0)$ mode, the locations and growth rate for the higher order $|\nu_1| > 1$ modes also change. As seen

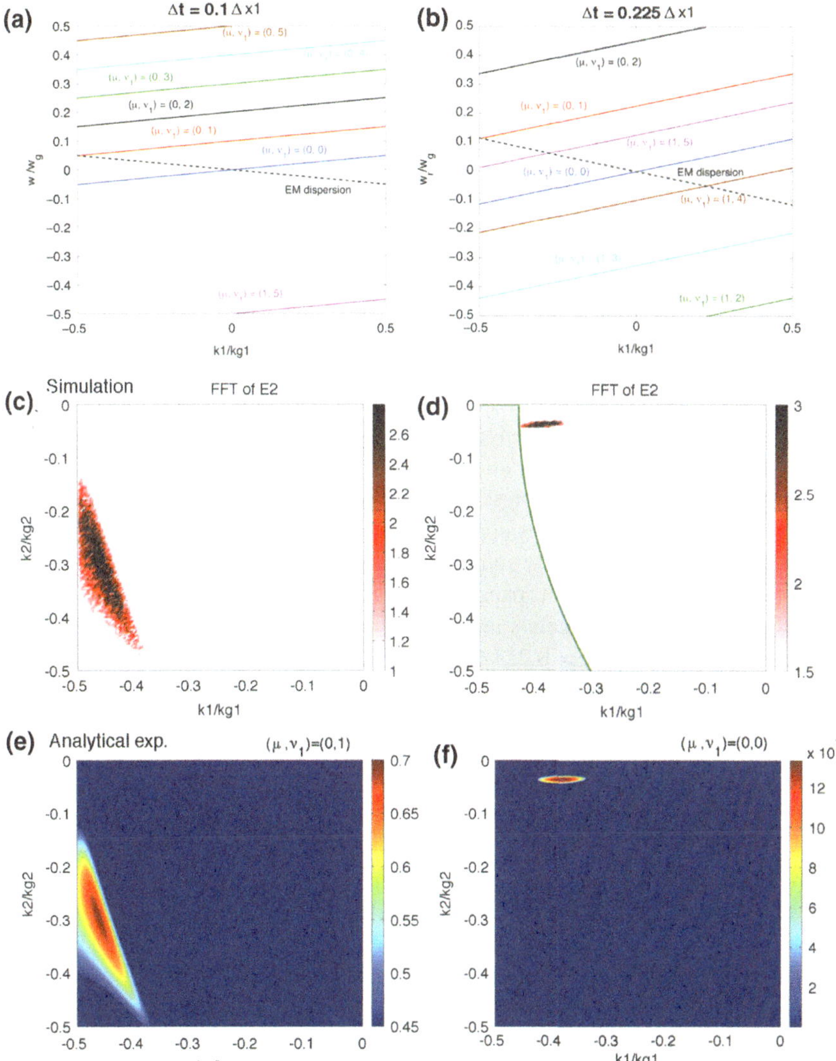

Fig. 5.10 **a** and **b** show the EM dispersion relation together with the beam resonance $\omega' - k_1' \beta = 0$, for $\Delta t = 0.1 \Delta x_1$ and $\Delta t = 0.225 \Delta x_1$ (and other parameters the same as listed in Table 5.1). **c** and **d** are the FFT of E_2 in the corresponding 2D simulations. The filter applied in order to observe the $(\mu, \nu_1) = (0, 0)$ mode is illustrated by the grey areas in (**d**). **e** and **f** are the corresponding analytical predictions by using the expression Eqs. 5.48 and 5.49 (Reprinted from Ref. [19]. Copyright with kind permission from Elsevier)

in Fig. 5.10a, the next beam resonance after $(\mu, \nu_1) = (0, \pm 1)$ is $(\mu, \nu_1) = (0, \pm 2)$ rather than $(\mu, \nu_1) = (1, \pm 1)$. It is easy to see that in this case the $(\mu, \nu_1) = (0, \pm 2)$ resonance line has no intersection with the EM mode in the fundamental Brillouin zone. The beam resonance line for an intermediate time step of $\Delta t = 0.225\Delta x_1$ are likewise shown in Fig. 5.10b. They show how gradually reducing the time step changes the NCI modes in the fundamental zone.

In the following, we will use these approaches to essentially eliminate the NCI in LWFA simulations in a Lorentz boosted frame, both of which involves the modeling of relativistically drifting plasma.

5.5.3 *LWFA Simulation in the Lorentz Boosted Frame with Spectral Solver*

We next present results from an LWFA boosted frame simulations in a nonlinear regime which is much more challenging. As shown in Ref. [21] excellent agreement between lab frame and boosted frame simulations are obtained in the linear regime using UPIC-EMMA when the fastest growing mode is filtered out. For simulation of nonlinear cases slight differences appear at higher γ_b. We revisit these simulations using strategies to systematically suppress the NCI modes.

In Fig. 5.11 we present results from a 2D LWFA simulations in the Lorentz booted frame. The simulation parameters are the same as in Ref. [21] with $\gamma_b = 28$. The time step used in [21] is $\Delta t = 0.221\Delta x_1$, which is about 1/2 of the 2D Courant limit for a standard spectral solver, and we used $\Delta t = 0.0525\Delta x_1$, which is about 1/8 the Courant limit in the new cases. In Ref. [21], for the case with $\gamma_b = 28$ we observed self-trapped particles that leads to a slightly different wake, as compared with both the $\gamma_b = 14$ case and the lab frame simulation. In Fig. 5.11c, d we show that by simply reducing the time step, we obtain results that are closer to the lab frame simulation. The plots of E_2 and plasma density also show subtle differences. The self-trapped electrons located at $k_0x_1 = 3.2 \times 10^4$ in the upper row of Fig. 5.11 are absent in the middle row. Note in the 2D OSIRIS lab frame simulation no self-trapped particles are observed. A consistent interpretation is also seen when transforming the boosted frame data back to the lab frame. In Fig. 5.11e we plot the line out of the E_1 wake field. This plot is the same as the second row of Fig. 8 in Ref. [21], except for the added data with $\Delta t = 0.0525\Delta x_1$ (black curves). It shows that better agreement to the lab frame and $\gamma_b = 14$ cases are found for the $\gamma_b = 28$ when the reduced time step are used (no self-trapped particles are seen in the lab frame and $\gamma_b = 14$ cases). For the simulation parameters of the reference simulation, the $(\mu, \nu_1) = (0, 0)$ mode is localized near $(\hat{k}_1, \hat{k}_2) = (0.24, 0.037)$ and the growth rate is $0.0341\omega_p$. For the reduced time step of $\Delta t/\Delta x_1 = 0.0525$, the $(\mu, \nu_1) = (0, 0)$ modes move to $(\hat{k}_1, \hat{k}_2) = (0.41, 0.023)$ and the growth rate is reduced to $0.0092\omega_p$.

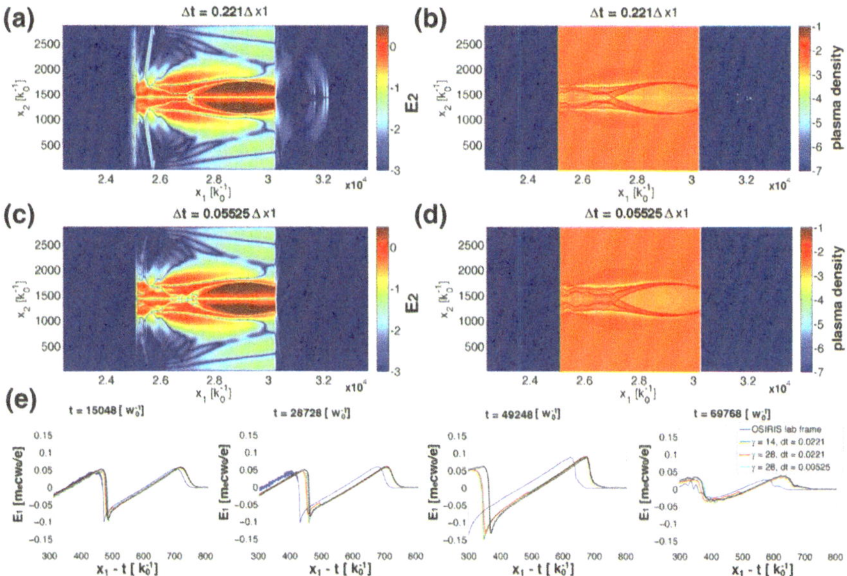

Fig. 5.11 **a** and **c** Shows the $\log_{10}|E_2|$ for the case $\Delta t = 0.221\Delta x_1$ and $\Delta t = 0.0525\Delta x_1$, respectively. **b** and **d** Shows the plasma density $\log_{10}|n|$ for the case $\Delta t = 0.221\Delta x_1$ and $\Delta t = 0.0525\Delta x_1$, respectively. **e** Shows the E_1 data comparison of the OSIRIS lab frame simulation, and the 3 UPIC-EMMA simulations

5.6 Conclusions

We derived a general multi-dimensional numerical dispersion relation for the relativistic plasma drift in the EM-PIC simulation that can include different choices in Maxwell solvers, differences between energy and momentum conserving field interpolation, differences between charge conserving and direct current deposition schemes, and the use of smoothing and low pass filters. Based on the dispersion relation we tried to understand the source of the instability and the structures of the dispersion relation. We confirmed that no instability occurs in 1D, and that in 2D and 3D a strong instability occurs due to the coupling between the Langmuir modes and transverse EM modes in the system. We can predict the pattern and growth rate of the instability for a particular simulation by solving the corresponding numerical dispersion relations. An asymptotic expression which permits rapid parameter scans for the ranges of unstable modes was derived. This asymptotic expression confirms and explains the existence of the 'magic time step'. These results are compared against simulation results using the EM-PIC code OSIRIS [10], as well as the UPIC-EMMA [11], and good agreements are obtained. Moreover, by plotting the intersection of the EM modes and beam resonances in (k_1, k_2, w_r) space the numerical instability patterns can be conveniently predicted. Maxwell Solvers, such as spectral solvers, that have EM waves with phase velocities greater than the speed of light are helpful to eliminate the instability due to the main beam resonance. We developed strategies for eliminating the NCI with an emphasis for a spectral solver. The principle idea is to ensure the unstable modes are far away from the physics of interest. We have used these strategies in UPIC-EMMA simulations of LWFA simulation in a Lorentz boosted frame and good agreements with lab frame simulations are obtained.

References

1. Vay JL (2007) Noninvariance of space- and time-scale ranges under a Lorentz transformation and the implications for the study of relativistic interactions. Phys Rev Lett 98:130405
2. Martins SF, Fonseca RA, Lu W et al (2010) Exploring laser-wakefield-accelerator regimes for near-term lasers using particle-in-cell simulation in Lorentz-boosted frames. Nat Phys 6(4):311–316
3. Jackson JD, Jackson JD (1962) Classical electrodynamics, vol 3. Wiley, New York
4. Martins SF, Fonseca RA, Silva LO et al (2010) Numerical simulations of laser wakefield accelerators in optimal Lorentz frames. Comput Phys Comm 181(5):869–875
5. Vay JL, Geddes CGR, Cormier-Michel E et al (2011) Numerical methods for instability mitigation in the modeling of laser wakefield accelerators in a Lorentz-boosted frame. J Comput Phys 230(15):5908–5929
6. Private discussion with Warren B Mori
7. Godfrey BB (1974) Numerical Cherenkov instabilities in electromagnetic particle codes. J Comput Phys 15(4):504–521
8. Godfrey BB (1975) Canonical momenta and numerical instabilities in particle codes. J Comput Phys 19(1):58–76
9. Esirkepov TZ (2001) Exact charge conservation scheme for Particle-in-Cell simulation with an arbitrary form-factor. Comput Phys Comm 135(2):144–153
10. Fonseca RA, Silva LO, Tsung FS et al (2002) OSIRIS: a three-dimensional, fully relativistic particle in cell code for modeling plasma based accelerators. Lecture notes in computer science, vol 2331, pp 342–351
11. Decyk VK (2007) UPIC: A framework for massively parallel particle-in-cell codes. Comput Phys Comm 177(1-2):95–97. Conference on computational physics (CCP 2006), Gyeongju, South Korea, Aug 29–Sep 01 2006
12. Langdon AB (1970) Effects of the spatial grid in simulation plasmas. J Comput Phys 6:247–267
13. Birdsall CK, Bruce LA (1985) Plasma physics via computer simulation. McGraw-Hill, New York
14. Yee KS (1966) Numerical solution of initial boundary value problems involving Maxwells equations in isotropic media. IEEE Trans Antennas Propag AP14(3):302
15. Kärkkäinen M, Gjonaj E, Lau T et al (2006) Low-dispersion wake field calculation tools. Proceedings of the 2006 international computational accelerator physics conference, Chamonix, France, 2006, pp 35–40
16. E T. High-order methods. In: Taflove A (ed) Advances in computational electrodynamics: the finite-difference time-domain methods. Artech House, Boston, 1998
17. Xu X, Yu P, Martins SF et al (2013) Numerical instability due to relativistic plasma drift in EM-PIC simulations. Comput Phys Commun 184(11):2503–2514
18. Godfrey BB, Vay JL (2013) Numerical stability of relativistic beam multidimensional PIC simulations employing the Esirkepov algorithm. J Comput Phys 248:33–46
19. Yu P, Xu X, Decyk VK et al (2015) Elimination of the numerical Cerenkov instability for spectral EM-PIC codes. Comput Phys Commun 192:32–47
20. Godfrey BB, Vay JL (2014) Suppressing the numerical Cherenkov instability in FDTD PIC codes. J Comput Phys 267:1–6
21. Yu P, Xu X, Decyk VK et al (2014) Modeling of laser wakefield acceleration in Lorentz boosted frame using EM-PIC code with spectral solver. J Comput Phys 266:124–138

Chapter 6
Summary

6.1 Concluding Remarks

Plasma based accelerators have made great progress in the past decade. To drive future compact X-ray FELs and colliders, the quality of the electron beam generated by current plasma based accelerators must be much improved. This needs a deeper understanding of the phase space dynamics of various injection methods and the beam transportation processes. On the other hand, plasma based accelerators does have a potential to produce electron beam with extremely good quality that may break the current limit of beam brightness imposed by traditional accelerator technology. There are many different new injection schemes in plasma based acceleration, and each own its unique injection process that may lead to special beam structure and quality. In this thesis, we focused on one kind of important injection schemes— ionization based injection method and studied its phase space dynamics in details. We then studied the important issue of beam transportation between plasma based accelerators and traditional accelerator components. Due to the very small β-function of electron beam from plasma accelerators, serious phase space mismatch can occur during the beam propagating between the plasma waves and the beam transportation components. We proposed to use longitudinally tailored plasma structure to match the beam between different stages. By combing high quality beam generation with the matching plasma structure, we presented a detailed study of a two-stage plasma based accelerator driven X-FEL. We also studied in great details the numerical noise in boosted frame simulations. A list of the major results are as follows:

1. The evolution of beam phase space in ionization injection in plasma wakefields is studied using theoretical analysis and PIC simulations. Two key phase mixing processes, namely longitudinal and transverse phase mixing, are identified to be responsible for the complex emittance dynamics that shows initial rapid growth followed by oscillation, decay, and eventually slow growth to saturation. An analytic theory is developed to include the effects of injection distance (time), acceleration distance, wakefield structure, and nonlinear space charge forces. Formulas for the emittance in the low and high charge regimes are presented and verified

© Springer Nature Singapore Pte Ltd. 2020

X. Xu, *Phase Space Dynamics in Plasma Based Wakefield Acceleration*,

Springer Theses, https://doi.org/10.1007/978-981-15-2381-6_6

through PIC simulations. A two-color laser ionization injection scheme for generating high quality beams is also examined through PIC simulations. In the end, the intrinsic phase space discretization in laser triggered ionization injection is analyzed.

2. The phase space matching between plasma acceleration stages and between plasma stages and traditional accelerator components is studied. Without proper matching, catastrophic emittance growth in the presence of finite energy spread may occur when the beam propagates through different stages and components due to the drastic differences of transverse focusing strength. A method using longitudinally tailored plasma structures as phase space matching components to properly guide the beam through stages is proposed. Theoretical analysis and particle-in-cell simulations are utilized to show clearly how these structures can work in four different scenarios.

3. A two-stage scheme of LWFA for driving a compact X-FEL is studied. In this scheme, an ultrashort 50 MeV electron beam with very high brightness is generated in a high density plasma using density downramp injection, and then the beam is sent through a low density plasma accelerator for further acceleration to reach a few GeV energy and to reduce the relative slice energy spread down to $\sim$0.4%. Plasma density matching structures are properly designed according to Chap. 3 to maintain the beam quality between the plasma stages. Finally, the X-FEL performance driven by such a high quality electron beam is studied using Genesis 1.3.

4. The numerical instability observed in electromagnetic particle-in-cell (EM-PIC) simulations with a plasma drifting with relativistic velocities is studied using both theory and computer simulations. We derive the numerical dispersion relation for a cold plasma drifting with a relativistic velocity, and find an instability attributed to the intersection between beam resonances and the electromagnetic modes in the drifting plasma. An asymptotic expression for the instability growth rate is obtained and has good agreements with the simulation results. Finally, we try to eliminate the numerical instability using spectral solver.

6.2 Future Work

Many important issues in plasma based acceleration that are relevant to the work described in this dissertation may be addressed in the future. A partial list is as follows:

1. The phase space dynamics of other injection methods can be studied using similar methods developed in this dissertation. These methods may have the potential to produce beams with even better quality than what has been achieved in the current known methods. Research along this line is still wide open.

2. For the high quality beam generated in plasma based accelerators, the electron density of the beam is typically very high, and this may lead to many new issues

due to the strong self space charge force. How these effects impact the beam quality during the injection and the acceleration process needs to be understood and quantified. The possibility of utilizing this effect is also an open question.

3. Due to the very low emittance and very high brightness of the electron beams generated in some plasma injection schemes, coherent Thomson scattering may occur by colliding the high quality beams with an optical undulator, such as a CO_2 laser pulse. The potential of such sources needs to be explored.

Appendix A
Derivation of the Emittance Evolution in the Acceleration Stage

In this section we derive some of the formulas discussed in Sect. 2.5.2. First, we consider the first three terms of Eq. 2.32. Integrate over E_z, we can get,

$$\int dE_z f_E(E_z) J_{11}^2 = \frac{\sqrt{\gamma_0}}{\Delta_E} \left[\sqrt{\frac{E_z}{\bar{\hat{s}}}} + \sqrt{\frac{E_z}{\bar{\hat{s}}}} \cos 2\phi + 4\sqrt{g}\,\mathrm{Si}\,(2\phi) \right] \Bigg|_{\bar{E}_z - \Delta_E/2}^{\bar{E}_z + \Delta_E/2}$$

$$\approx \frac{\sqrt{\gamma_0}}{\Delta_E} \left[\sqrt{\frac{E_z}{\bar{\hat{s}}}} + \frac{\pi}{2} - \frac{E_z}{4\sqrt{g}\bar{\hat{s}}} \sin 2\phi \right] \Bigg|_{\bar{E}_z - \Delta_E/2}^{\bar{E}_z + \Delta_E/2}$$

$$\approx \frac{1}{2}\sqrt{\frac{\gamma_0}{\bar{E}_z\bar{\hat{s}}}} + \frac{\bar{E}_z}{2\bar{\hat{s}}\Delta_E}\sqrt{\frac{\gamma_0}{g}} \sin\left(\sqrt{\frac{g\bar{\hat{s}}}{\bar{E}_z}}\frac{\Delta_E}{\bar{E}_z} \right) \cos 2\phi$$

$$\int dE_z f_E(E_z) J_{12}^2 \approx \frac{1}{2g\sqrt{\bar{E}_z\bar{\hat{s}}\gamma_0}} - \frac{\bar{E}_z}{2\bar{\hat{s}}g^{3/2}\Delta_E\sqrt{\gamma_0}} \sin\left(\sqrt{\frac{g\bar{\hat{s}}}{\bar{E}_z}}\frac{\Delta_E}{\bar{E}_z} \right) \cos 2\phi$$

$$\int dE_z f_E(E_z) J_{11}J_{21} = -\frac{\sqrt{\gamma_0}}{\Delta_E} \left[2g\sqrt{E_z\bar{\hat{s}}}\cos 2\phi + \frac{\sqrt{g}E_z}{2}\sin 2\phi + 8g^{3/2}\bar{\hat{s}}\,\mathrm{Si}\,(2\phi) \right] \Bigg|_{\bar{E}_z - \Delta_E/2}^{\bar{E}_z + \Delta_E/2}$$

$$\approx -\frac{\sqrt{\gamma_0}}{\Delta_E} \left[4g^{3/2}\pi\bar{\hat{s}} + \frac{1}{4}\sqrt{\frac{E_z^3}{\bar{\hat{s}}}}\cos 2\phi \right] \Bigg|_{\bar{E}_z - \Delta_E/2}^{\bar{E}_z + \Delta_E/2}$$

$$\approx -\frac{1}{2}\sqrt{\frac{\bar{E}_z\gamma_0}{\bar{\hat{s}}}} \frac{\bar{E}_z}{\Delta_E}\sin\left(\sqrt{\frac{g\bar{\hat{s}}}{\bar{E}_z}}\frac{\Delta_E}{\bar{E}_z} \right) \sin 2\phi$$

$$\int dE_z f_E(E_z) J_{12}J_{22} \approx \frac{1}{2g}\sqrt{\frac{\bar{E}_z}{\bar{\hat{s}}\gamma_0}} \frac{\bar{E}_z}{\Delta_E}\sin\left(\sqrt{\frac{g\bar{\hat{s}}}{\bar{E}_z}}\frac{\Delta_E}{\bar{E}_z} \right) \sin 2\phi$$

© Springer Nature Singapore Pte Ltd. 2020
X. Xu, *Phase Space Dynamics in Plasma Based Wakefield Acceleration*,
Springer Theses, https://doi.org/10.1007/978-981-15-2381-6

$$\int dE_z f_E(E_z) J_{21}^2 = \frac{g\sqrt{\gamma_0}}{3\Delta_E}\left[E_z^{3/2}\sqrt{\bar{\bar{s}}} + 32 g^{3/2}\bar{\bar{s}}^2 \mathrm{Si}\,(2\phi) - 8 g\bar{\bar{s}}^{3/2}\sqrt{E_z}\left(\frac{E_z}{8g\bar{\bar{s}}} - 1\right)\cos 2\phi \right.$$
$$\left. \left. + 2\sqrt{g}\, E_z \bar{\bar{s}}\sin 2\phi \right]\right|_{\bar{E}_z-\Delta_E/2}^{\bar{E}_z+\Delta_E/2}$$

$$\approx \frac{g\sqrt{\gamma_0}}{3\Delta_E}\left[E_z^{3/2}\sqrt{\bar{\bar{s}}} + 16 g^{3/2}\pi\bar{\bar{s}}^2 + \frac{3}{4\sqrt{g}} E_z^2\sin 2\phi \right]\Bigg|_{\bar{E}_z-\Delta_E/2}^{\bar{E}_z+\Delta_E/2}$$

$$\approx \frac{g\sqrt{\bar{E}_z\bar{\bar{s}}\gamma_0}}{2} - \frac{\sqrt{g\gamma_0}\bar{E}_z}{2}\frac{\bar{E}_z}{\Delta_E}\sin\left(\sqrt{\frac{g\bar{\bar{s}}}{\bar{E}_z}}\frac{\Delta_E}{\bar{E}_z}\right)\cos 2\phi$$

$$\int dE_z f_E(E_z) J_{22}^2 \approx \frac{1}{2}\sqrt{\frac{\bar{E}_z\bar{\bar{s}}}{\gamma_0}} + \frac{\bar{E}_z}{2\sqrt{g\gamma_0}}\frac{\bar{E}_z}{\Delta_E}\sin\left(\sqrt{\frac{g\bar{\bar{s}}}{\bar{E}_z}}\frac{\Delta_E}{\bar{E}_z}\right)\cos 2\phi \tag{A.1}$$

where $f_E(E_z) = \mathrm{rect}\left(\frac{E_z-\bar{E}_z}{\Delta_E}\right)$ and $\bar{E}_z \gg \Delta_E$ are used.

Some integration formulas are listed as follows,

$$\int dx \sin\left(\sqrt{\frac{1}{x}}\right) = \sqrt{x}\cos\left(\sqrt{\frac{1}{x}}\right) + x\sin\left(\sqrt{\frac{1}{x}}\right) + \mathrm{Si}\left(\sqrt{\frac{1}{x}}\right) \tag{A.2}$$

$$\int dx\sqrt{\frac{1}{x}}\cos\left(\sqrt{\frac{1}{x}}\right) = 2\sqrt{x}\cos\left(\sqrt{\frac{1}{x}}\right) + 2\mathrm{Si}\left(\sqrt{\frac{1}{x}}\right) \tag{A.3}$$

$$\int dx\sqrt{x}\cos\left(\sqrt{\frac{1}{x}}\right) = \frac{1}{3}\left[\sqrt{x}(2x-1)\cos\left(\sqrt{\frac{1}{x}}\right) - x\sin\left(\sqrt{\frac{1}{x}}\right) - \mathrm{Si}\left(\sqrt{\frac{1}{x}}\right) \right] \tag{A.4}$$

where $\mathrm{Si}(x)$ is the sine integral function defined as $\mathrm{Si}(x) = \int_0^x dt\,\frac{\sin t}{t}$, and when $x \gg 1$,

$$\mathrm{Si}(x) \approx \frac{\pi}{2} - \frac{\cos x}{x}\sum_{n=0}^{+\infty}\frac{(-1)^n(2n)!}{x^{2n}} - \frac{\sin x}{x}\sum_{n=0}^{+\infty}\frac{(-1)^n(2n+1)!}{x^{2n+1}} \tag{A.5}$$

After some algebra calculations, the first three terms of Eq. 2.32 can be obtained as,

$$\sigma_{x0}^4 S_1 + \sigma_{x0}^2\sigma_{px0}^2 S_2 + \sigma_{px0}^4 S_3$$
$$= \frac{\left(g\gamma_0\sigma_{x0}^2 + \sigma_{px0}^2\right)^2}{4g\gamma_0} - \frac{\bar{E}_z}{4g^2\gamma_0\bar{\bar{s}}}\left(\frac{\bar{E}_z}{\Delta_E}\right)^2\left(g\gamma_0\sigma_{x0}^2 - \sigma_{px0}^2\right)^2\sin^2\left(\sqrt{\frac{g\bar{\bar{s}}}{\bar{E}_z}}\frac{\Delta_E}{\bar{E}_z}\right) \tag{A.6}$$

Next, we discuss the coefficient of Δ^2 in Eq. 2.32 which are,

$$
S_4 \approx \frac{g\gamma_0}{\bar{\hat{s}}^2} \left[\frac{1}{8} - \frac{\cos\left(\sqrt{\frac{g\bar{\hat{s}}}{\bar{E}_z}}\frac{\Delta_E}{\bar{E}_z}\right)\cos 2\phi}{4} + \frac{\sin^2\left(\sqrt{\frac{g\bar{\hat{s}}}{\bar{E}_z}}\frac{\Delta_E}{\bar{E}_z}\right)}{8} + \frac{\bar{E}_z}{\Delta_E}\sin\left(\sqrt{\frac{g\bar{\hat{s}}}{\bar{E}_z}}\frac{\Delta_E}{\bar{E}_z}\right)\sin 2\phi \right.
$$

$$
\left. + 2\left(\frac{\bar{E}_z}{\Delta_E}\right)^2 \sin^2\left(\sqrt{\frac{g\bar{\hat{s}}}{\bar{E}_z}}\frac{\Delta_E}{\bar{E}_z}\right) \right]
$$

$$
\approx \frac{2g\gamma_0}{\bar{\hat{s}}^2}\left(\frac{\bar{E}_z}{\Delta_E}\right)^2 \sin^2\left(\sqrt{\frac{g\bar{\hat{s}}}{\bar{E}_z}}\frac{\Delta_E}{\bar{E}_z}\right)
$$

$$
S_5 \approx -\frac{1}{\bar{\hat{s}}^2}\left[2\left(\frac{\bar{E}_z}{\Delta_E}\right)^2 - 2\left(\frac{\bar{E}_z}{\Delta_E}\right)^2\cos\left(2\sqrt{\frac{g\bar{\hat{s}}}{\bar{E}_z}}\frac{\Delta_E}{\bar{E}_z}\right) - \frac{1}{8} - \frac{1}{8}\cos\left(2\sqrt{\frac{g\bar{\hat{s}}}{\bar{E}_z}}\frac{\Delta_E}{\bar{E}_z}\right) \right]
$$

$$
\approx -\frac{4}{\bar{\hat{s}}^2}\left(\frac{\bar{E}_z}{\Delta_E}\right)^2 \sin^2\left(\sqrt{\frac{g\bar{\hat{s}}}{\bar{E}_z}}\frac{\Delta_E}{\bar{E}_z}\right)
$$

$$
S_6 \approx \frac{1}{g\bar{\hat{s}}^2\gamma_0}\left[\frac{1}{8} + \frac{\cos\left(\sqrt{\frac{g\bar{\hat{s}}}{\bar{E}_z}}\frac{\Delta_E}{\bar{E}_z}\right)\cos 2\phi}{4} + \frac{\sin^2\left(\sqrt{\frac{g\bar{\hat{s}}}{\bar{E}_z}}\frac{\Delta_E}{\bar{E}_z}\right)}{8} \right.
$$

$$
\left. - \frac{\bar{E}_z}{\Delta_E}\sin\left(\sqrt{\frac{g\bar{\hat{s}}}{\bar{E}_z}}\frac{\Delta_E}{\bar{E}_z}\right)\sin 2\phi + \left(\frac{\bar{E}_z}{\Delta_E}\right)^2 2\sin^2\left(\sqrt{\frac{g\bar{\hat{s}}}{\bar{E}_z}}\frac{\Delta_E}{\bar{E}_z}\right) \right]
$$

$$
\approx \frac{2}{g\bar{\hat{s}}^2\gamma_0}\left(\frac{\bar{E}_z}{\Delta_E}\right)^2 \sin^2\left(\sqrt{\frac{g\bar{\hat{s}}}{\bar{E}_z}}\frac{\Delta_E}{\bar{E}_z}\right) \tag{A.7}
$$

where high order terms are neglected. Then, Eq. 2.32 can be obtained.

Appendix B
Interpolation Tensor and Finite Difference Operator

In this section we will write out the explicit expressions for the interpolation tensors $\overleftrightarrow{S}$ for the fields and the currents used in Chap. 5. For a momentum conserving scheme in 3D the interpolation tensor for the EM field after the Fourier transform can be expressed as:

$$S_{E1} = s_{l,1}s_{l,2}s_{l,3}\eta_1 \quad S_{E2} = s_{l,1}s_{l,2}s_{l,3}\eta_2 \quad S_{E3} = s_{l,1}s_{l,2}s_{l,3}\eta_3$$
$$S_{B1} = \cos(\omega\Delta t/2)s_{l,1}s_{l,2}s_{l,3}\eta_2\eta_3 \quad S_{B2} = \cos(\omega\Delta t/2)s_{l,1}s_{l,2}s_{l,3}\eta_1\eta_3$$
$$S_{B3} = \cos(\omega\Delta t/2)s_{l,1}s_{l,2}s_{l,3}\eta_1\eta_2 \tag{B.1}$$

where

$$s_{l,i} = \left(\frac{\sin(k_i\Delta x_i/2)}{k_i\Delta x_i/2}\right)^{l+1} \tag{B.2}$$

and $\eta_i = \zeta^{\nu_i}$, $\zeta = -1$ when the EM field has a half-grid offset in the $\hat{i}$ direction, and $\zeta = 1$ when it is defined at grid point. l refers to the order ($l = 1$ is area weighting or linear interpolation for the charge). While for the energy conserving scheme, we have

$$S_{E1} = s_{l-1,1}s_{l,2}s_{l,3}\eta_1 \quad S_{E2} = s_{l,1}s_{l-1,2}s_{l,3}\eta_2 \quad S_{E3} = s_{l,1}s_{l,2}s_{l-1,3}\eta_3$$
$$S_{B1} = \cos(\omega\Delta t/2)s_{l,1}s_{l-1,2}s_{l-1,3}\eta_2\eta_3 \quad S_{B2} = \cos(\omega\Delta t/2)s_{l-1,1}s_{l,2}s_{l-1,3}\eta_1\eta_3$$
$$S_{B3} = \cos(\omega\Delta t/2)s_{l-1,1}s_{l-1,2}s_{l,3}\eta_1\eta_2 \tag{B.3}$$

The space finite difference operator for the Yee solver is:

$$[k]_i = \frac{\sin(k_i\Delta x_i/2)}{\Delta x_i/2} \tag{B.4}$$

and is the same for electric and magnetic field.

© Springer Nature Singapore Pte Ltd. 2020
X. Xu, *Phase Space Dynamics in Plasma Based Wakefield Acceleration*,
Springer Theses, https://doi.org/10.1007/978-981-15-2381-6

In Karkkainen solver, the space finite difference operator for the magnetic field $[k]_{Bi}$ is the same as Eq. B.4, while for the electric field

$$[k]_{Ei} = c_i \frac{\sin(k_i \Delta x_i/2)}{\Delta x_i/2} \tag{B.5}$$

where

$$c_1 = \theta_1 + 2\theta_2\{\cos(k_2\Delta x_2) + \cos(k_3\Delta x_3)\} + 4\theta_3 \cos(k_2\Delta x_2)\cos(k_3\Delta x_3)$$
$$c_2 = \theta_1 + 2\theta_2\{\cos(k_3\Delta x_3) + \cos(k_1\Delta x_1)\} + 4\theta_3 \cos(k_3\Delta x_3)\cos(k_1\Delta x_1)$$
$$c_3 = \theta_1 + 2\theta_2\{\cos(k_1\Delta x_1) + \cos(k_2\Delta x_2)\} + 4\theta_3 \cos(k_1\Delta x_1)\cos(k_2\Delta x_2) \tag{B.6}$$

and

$$\theta_1 = 7/12 \quad \theta_2 = 1/12 \quad \theta_3 = 1/48 \tag{B.7}$$

are the tunable parameters for the Karkkainen solver [1]. The space finite difference operator for the spectral solver is

$$[k]_i = k_i \tag{B.8}$$

The time finite difference operator for the Yee, Karkkainen, and spectral solvers are the same

$$[\omega] = \frac{\sin(\omega\Delta t/2)}{\Delta t/2} \tag{B.9}$$

With respect to the current interpolation, the current interpolation tensor is approximately:

$$S_{j1} = s_{l-1,1}s_{l,2}s_{l,3}\eta_1 \quad S_{j2} = s_{l,1}s_{l-1,2}s_{l,3}\eta_2 \quad S_{j3} = s_{l,1}s_{l,2}s_{l-1,3}\eta_3 \tag{B.10}$$

We note that the expressions for $\overleftrightarrow{S_j}$ are for the charge conserving current deposition scheme of vanishing time step $\Delta t \to 0$. Nonetheless, one can obtain the corresponding $\overleftrightarrow{\epsilon}$ for the exact charge conserving scheme by replacing Eq. 5.5 with Eqs. 19 and 23 of Ref. [2] and then Fourier analyzing the expression. This was done in Ref. [3]. We note that using the more exact form in the dispersion relation the instability growth rates change by only a few percent. In addition, it does not change the presence of the optimal time step, i.e, the optimal time step arises from the staggering of the fields on the mesh and not the choice of the current deposit.

Meanwhile, when the current is directly deposited (as is done in the UPIC framework), the current interpolation functions are,

$$S_{j1} = s_{l,1}s_{l,2}s_{l,3}\eta_1 \quad S_{j2} = s_{l,1}s_{l,2}s_{l,3}\eta_2 \quad S_{j3} = s_{l,1}s_{l,2}s_{l,3}\eta_3 \tag{B.11}$$

Note that in the spectral algorithm, at each time step the longitudinal electric field is directly obtained by solving Gauss's law which effectively ensures charge conservation.

References

1. Kärkkäinen M, Gjonaj E, Lau T et al (2006) Low-dispersion wake field calculation tools. In: Proceedings of the 2006 international computational accelerator physics conference, Chamonix, France, pp 35–40
2. Esirkepov TZ (2001) Exact charge conservation scheme for Particle-in-Cell simulation with an arbitrary form-factor. Comput Phys Commun 135(2):144–153
3. Godfrey BB, Vay JL (2013) Numerical stability of relativistic beam multidimensional PIC simulations employing the Esirkepov algorithm. J Comput Phys 248:33–46

The manufacturer's authorised representative in the EU is Springer
Nature Customer Service Centre GmbH, Europaplatz 3, 69115 Heidelberg,
Germany. If you have any concerns regarding our products, please
contact ProductSafety@springernature.com

Printed and bound by CPI Group (UK) Ltd, Croydon, CR0 4YY

28/11/2025

02007727-0006